*James M. McCloskey*
*Editor*

# A Guide to Docutek, Inc.'s ERes Software: A Way to Manage Electronic Reserves

*A Guide to Docutek, Inc.'s ERes Software: A Way to Manage Electronic Reserves* has been co-published simultaneously as *Journal of Interlibrary Loan, Document Delivery & Electronic Reserve*, Volume 15, Number 1 2004.

*Pre-publication*
*REVIEWS,*
*COMMENTARIES,*
*EVALUATIONS . . .*

"INFORMATIVE. . . . This collection discusses very specific technical considerations such as, 'Do I need a dedicated server to run ERes?' as well as tackling broader policy issues such as copyright and fair use for electronic reserves. Veteran e-reserves practitioners and newcomers alike will find much to be used in this multifaceted volume that COVERS ALL PHASES OF DEVELOPING AN ELECTRONIC RESERVES SERVICE–from pilot projects, migration, and workflow to integrating e-reserves with other campus tools like courseware. This is not just a timeline of a product's evolution. It is a timely snapshot of current issues surrounding this mission-critical academic library service."

**Leah G. McGinnis, MSLS**
*Head*
*R. B. House Undergraduate Library*
*University of North Carolina at Chapel Hill*

*More pre-publication*
*REVIEWS, COMMENTARIES, EVALUATIONS . . .*

"This is A PRACTICAL BOOK that that provides both a starting place and an outline of what needs to be done in order to choose a system and to integrate electronic reserves into an academic library environment. It takes into consideration that we do not all start from the same place but that we all start from the premise that we want to provide the best possible service to our patrons. It is no surprise that The Haworth Press, Inc., has brought out a monograph on this timely and interesting subject. What may be a surprise to readers is the wide diversity of information provided in this volume. Professionals who have 'been there and done that' have written of their experiences so that we may all learn from them and apply that which best applies to our own situations. WE DID A GREAT DEAL OF RESEARCH BEFORE WE CHOSE AN ELECTRONIC RESERVES SYSTEM AT MY INSTITUTION. IF THIS BOOK HAD EXISTED AT THAT TIME, THE TASK WOULD HAVE BEEN A GREAT DEAL EASIER."

**June L. DeWeese, MLS**
*Head of Access Services*
*Ellis Library*
*University of Missouri-Columbia*

"COVERS ALL ASPECTS OF DOCU-TEK'S ERes SYSTEM. . . . In this one volume, all of the information needed about Docutek ERes software, its implementation, and its shortcomings are discussed in a manner that will inform both the current user and those considering an electronic reserves product. Especially interesting are the connections with courseware and the incorporation of ERes into other areas of faculty use of online resources."

**Joyce Rumery, MLS**
*Interim Director*
*Fogler Library*
*University of Maine*

# A Guide to Docutek, Inc.'s ERes Software: A Way to Manage Electronic Reserves

*A Guide to Docutek, Inc.'s ERes Software: A Way to Manage Electronic Reserves* has been co-published simultaneously as *Journal of Interlibrary Loan, Document Delivery & Electronic Reserve*, Volume 15, Number 1 2004.

# The *Journal of Interlibrary Loan, Document Delivery & Electronic Reserve*™ Monographic "Separates"

(formerly the *Journal of Interlibrary Loan, Document Delivery & Information Supply*\*\* Volumes 4-14 and the *Journal of Interlibrary Loan & Information Supply*\* Volumes 1-3)

For information on previous issues of the *Journal of Interlibrary Loan & Information Supply* series and the *Journal of Interlibrary Loan, Document Delivery & Information Supply* series, edited by Leslie R. Morris, please contact: The Haworth Press, Inc., 10 Alice Street, Binghamton, NY 13904-1580 USA.

Below is a list of "separates," which in serials librarianship means a special issue simultaneously published as a journal issue or double-issue *and* as a "separate" hardbound monograph. (This is a format which we also call a "DocuSerial.")

"Separates" are published because specialized libraries or professionals may wish to purchase a specific thematic issue by itself in a format which can be separately cataloged and shelved, as opposed to purchasing the journal on an on-going basis. Faculty members may also more easily consider a "separate" for classroom adoption.

"Separates" are carefully classified separately with the major book jobbers so that the journal tie-in can be noted on new book order slips to avoid duplicate purchasing.

You may wish to visit Haworth's Website at . . .

## http://www.HaworthPress.com

. . . to search our online catalog for complete tables of contents of these separates and related publications.

You may also call 1-800-HAWORTH (outside US/Canada: 607-722-5857), or Fax 1-800-895-0582 (outside US/Canada: 607-771-0012), or e-mail at:

## docdelivery@haworthpress.com

---

*A Guide to Docutek, Inc.'s ERes Software: A Way to Manage Electronic Reserves,* edited by James M. McCloskey (Vol. 15, No. 1, 2004). *Presents first-hand insights and information on how several academic libraries selected and implemented one of the leading electronic reserve systems.*

*Legal Solutions in Electronic Reserves and the Electronic Delivery of Interlibrary Loan,*\*\* by Janet Brennan Croft (Vol. 14, No. 3, 2004). *Guides librarians through the process of developing legal policies for their electronic resources to protect patrons' rights and to avoid copyright infringement.*

*Electronic Reserve: A Manual and Guide for Library Staff Members,*\*\* by Lori Driscoll (Vol. 14, No. 1, 2003). *A comprehensive guide to establishing and maintaining effective electronic reserve services.*

*Interlibrary Loan and Document Delivery in the Larger Academic Library: A Guide for University, Research, and Larger Public Libraries,*\*\* by Lee Andrew Hilyer (Vol. 13, No. 1/2, 2002). *A concise but thorough introductory guide to the daily operation of an interlibrary loan department.*

*Ariel: Internet Transmission Software for Document Delivery,*\*\* edited by Gary Ives (Vol. 10, No. 4, 2000). *"Useful . . . Instructive. Answers are provided to questions continually asked by other Ariel users as well as potential users." (Marilyn C. Grush, MLS, Coordinator, Interlibrary Loan/Document Delivery, University of Delaware)*

*Information Delivery in the 21st Century: Proceedings of the Fourth International Conference on Fee-Based Information Services in Libraries,*\*\* edited by Suzanne M. Ward, Yem S. Fong, and Tammy Nickelson Dearie (Vol. 10, No. 1, 1999). *"This book is an excellent overview of the issues and realities of fee-based information services in academic and public libraries. . . . It is especially insightful for public libraries considering fee-based services. . . . An excellent addition to any library's collection." (Kathy Gillespie Tomajko, MLn, BS, Department Head, Reference Services, Georgia Institute of Technology Library and Information Center)*

***The Economics of Access versus Ownership: The Costs and Benefits of Access to Scholarly Articles via Interlibrary Loan and Journal Subscriptions,*** by Bruce R. Kingma, PhD (Vol. 6, No. 3, 1996). *"Presents a well-constructed and well-described study and a consequent set of conclusions about the cooperative economics of borrowing versus owning library journal subscriptions. . . . A well-done and much needed book." (Catholic Library World)*

***Information Brokers: Case Studies of Successful Ventures,*** by Alice Jane Holland Johnson, MLS (Vol. 5, No. 2, 1995). *"The insights in this compilation give practical overviews that are applicable to information professionals interested in becoming information brokers, starting their own brokerages, or adding this function to their existing library service." (Journal of Interlibrary Loan, Document Delivery & Information Supply)*

***Interlibrary Loan of Alternative Format Materials: A Balanced Sourcebook,**** edited by Bruce S. Massis, MLS, MA and Winnie Vitzansky (Vol. 3, No. 1/2, 1993). *"Essential for interlibrary loan departments serving blind or visually handicapped patrons. . . . An enlightening survey of the state of the art in international lending of nonprint library materials." (Information Technology and Libraries)*

# A Guide to Docutek, Inc.'s ERes Software: A Way to Manage Electronic Reserves

James M. McCloskey

Editor

*A Guide to Docutek, Inc.'s ERes Software: A Way to Manage Electronic Reserves* has been co-published simultaneously as *Journal of Interlibrary Loan, Document Delivery & Electronic Reserve*, Volume 15, Number 1 2004.

The Haworth Information Press®
An Imprint of The Haworth Press, Inc.

New York • London • Victoria (AU)
**www.HaworthPress.com**

Published by

The Haworth Information Press®, 10 Alice Street, Binghamton, NY 13904-1580 USA

The Haworth Information Press® is an imprint of The Haworth Press, Inc., 10 Alice Street, Binghamton, NY 13904-1580 USA.

*A Guide to Docutek, Inc.'s ERes Software: A Way to Manage Electronic Reserves* has been co-published simultaneously as *Journal of Interlibrary Loan, Document Delivery & Electronic Reserve*™, Volume 15, Number 1 2004.

Cover design by Lora Wiggins.

**Library of Congress Cataloging-in-Publication Data**

A guide to Docutek, Inc.'s ERes software : a way to manage electronic reserves / James M. McCloskey, editor.
    p. cm.
    "Co-published simultaneously as Journal of interlibrary loan, document delivery & electronic reserve, volume 15, number 1, 2004."
    Includes index.
    ISBN 0-7890-2782-8 (alk. paper) – ISBN 0-7890-2783-6 (pbk. : alk. paper)
    1. Docutek ERes. 2. Electronic reserve collections in libraries–Management–Computer programs. 3. Academic libraries–Electronic reserve collections–Management–Computer programs. I. McCloskey, James M. II. Journal of interlibrary loan, document delivery & electronic reserve.
Z692.R47 G85 2005
025.04–dc22
                                       2004021155

# Indexing, Abstracting & Website/Internet Coverage

This section provides you with a list of major indexing & abstracting services and other tools for bibliographic access. That is to say, each service began covering this periodical during the year noted in the right column. Most Websites which are listed below have indicated that they will either post, disseminate, compile, archive, cite or alert their own Website users with research-based content from this work. (This list is as current as the copyright date of this publication.)

Abstracting, Website/Indexing Coverage . . . . . . . . Year When Coverage Began

- *Academic Abstracts/CD-ROM* . . . . . . . . . . . . . . . . . . . . . . . . . . . . . . . . . . . . . . . . 1995

- *Academic Search: database of 2,000 selected academic serials, updated monthly: EBSCO Publishing* . . . . . . . . . . . . . . . . . . . . . . . . . . . . . . . . . . . . . . . . 1995

- *Academic Search Elite (EBSCO)* . . . . . . . . . . . . . . . . . . . . . . . . . . . . . . . . . . . . . 1995

- *Academic Search Premier (EBSCO)* <br> *<http://www.epnet.com/academic/acasearchprem.asp>* . . . . . . . . . . . . . . . . . . . . . . 1995

- *Business Source Corporate: coverage of nearly 3,350 quality magazines and journals; designed to meet the diverse information needs of corporations; EBSCO Publishing* <br> *<http://www.epnet.com/corporate/bsourcecorp.asp>* . . . . . . . . . . . . . . . . . . . . . . . 1995

- *Chartered Institute of Library and Information Professionals (CILIP) Group Newsletter, supplement to Health Libraries Review, official journal of LA HLG. Published quarterly by Blackwell Science* <br> *<http://www.blackwell-science.com/hlr/newsletter/>* . . . . . . . . . . . . . . . . . . . . . . . . 1998

- *Computer and Information Systems Abstracts <http://www.csa.com>* . . . . . . . . . . . . 2004

- *Current Cites [Digital Libraries] [Electronic Publishing] [Multimedia & Hypermedia] [Networks & Networking] [General]* <br> *<http://sunsite.berkeley.edu/CurrentCites/>* . . . . . . . . . . . . . . . . . . . . . . . . . . . . . 2000

- *Current Index to Journals in Education* . . . . . . . . . . . . . . . . . . . . . . . . . . . . . . . . 2001

- *FRANCIS. INIST/CNRS <http://www.inist.fr>* . . . . . . . . . . . . . . . . . . . . . . . . . . . . 1999

- *IBZ International Bibliography of Periodical Literature* <br> *<http://www.saur.de>* . . . . . . . . . . . . . . . . . . . . . . . . . . . . . . . . . . . . . . . . . . . . . . 1995

(continued)

- *Index Guide to College Journals (core list compiled by integrating 48 indexes frequently used to support undergraduate programs in small to medium sized libraries)* . . . . . . . . . . . . . . . . . . . . . . . . . . . . . . . . . . . . . . .1999

- *Index to Periodical Articles Related to Law* <http://www.utexas.edu> . . . . . . . . . . . . 1991

- *Information Reports & Bibliographies* . . . . . . . . . . . . . . . . . . . . . . . . . . . . . . . . . .1991

- *Information Science & Technology Abstracts: indexes journal articles from more than 450 publications as well as books, research reports, and conference proceedings; EBSCO Publishing* <http://www.epnet.com>. . . . . . . . . . . . . . . . . .1993

- *Informed Librarian, The. For more information visit us at:* <http://www.informedlibrarian.com> . . . . . . . . . . . . . . . . . . . . . . . . . . .1993

- *INSPEC is the leading English-language bibliographic information service providing access to the world's scientific & technical literature in physics, electrical engineering, electronics, communications, control engineering, computers & computing, and information technology* <http://www.iee.org.uk/publish/> . . . . . . . . . . . . . . . . . . . . . . . . . . . . . . . . . . . . .1993

- *Internationale Bibliographie der geistes- und sozialwissenschaftlichen Zeitschriftenliteratur . . . See IBZ.* . . . . . . . . . . . . . . . . . . . . . . . . . . . . . . . . . . . .1995

- *Journal of Academic Librarianship: Guide to Professional Literature, The* . . . . . . . . . . . . . . . . . . . . . . . . . . . . . . . . . . . . . . . . . . . . . . . . . . . . . . .1997

- *Konyvtari Figyelo (Library Review).* . . . . . . . . . . . . . . . . . . . . . . . . . . . . . . . . . . .1995

- *Library & Information Science Abstracts (LISA)* <http://www.csa.com> . . . . . . . . . .1991

- *Library and Information Science Annual (LISCA)* <http://www.lu.com>. . . . . . . . . . .1997

- *Library Literature & Information Science* <http://www.hwwilson.com> . . . . . . . . . .1991

- *Library Reference Center: Comprised of indexing and abstracting from more than 90 important library trade magazines and journals; EBSCO Publishing* <http://www.epnet.com> . . . . . . . . . . . . . . . . . . . . . . . . . . . . . . . . . . . . . . . . . . . .2001

- *MasterFILE: updated database from EBSCO Publishing* . . . . . . . . . . . . . . . . . . . .1995

- *MasterFILE Elite: coverage of nearly 1,200 periodicals covering general reference, business, health, education, general science, multi-cultural issues and much more; EBSCO Publishing* <http://www.epnet.com/government/mfelite.asp> . . . . . . .1995

- *MasterFILE Premier: coverage of more than 1,950 periodicals covering general reference, business, health, education, general science, multi-cultural issues and much more; EBSCO Publishing* <http://www.epnet.com/government/mfpremier.asp> . . . . . . . . . . . . . . . . . . . . .1995

- *MasterFILE Select: coverage of nearly 770 periodicals covering general reference, business, health, education, general science, multi-cultural issues and much more; EBSCO Publishing* <http://www.epnet.com/government/mfselect.asp> . . . . . . . . . . . . . . . . . . . . . .1995

- *OCLC ArticleFirst* <http://www.oclc.org/services/databases/> . . . . . . . . . . . . . . . . . .2003

- *OCLC ContentsFirst* <http://www.oclc.org/services/databases/> . . . . . . . . . . . . . . . .2003

(continued)

- *PASCAL, c/o Institut de l'Information Scientifique et Technique. Cross-disciplinary electronic database covering the fields of science, technology & medicine. Also available on CD-ROM, and can generate customized retrospective searches <http://www.inist.fr>* . . . . . . . . . . . . . . . . . . . . . . . . . . . . . . . . . . . . . . . . 1996

- *Referativnyi Zhurnal (Abstracts Journal of the All-Russian Institute of Scientific and Technical Information–in Russian)* . . . . . . . . . . . . . . . 1991

- *Sage Public Administration Abstracts (SPAA)* . . . . . . . . . . . . . . . . . . . . . . . . . . . 1991

- *SwetsWise <http://www.swets.com>* . . . . . . . . . . . . . . . . . . . . . . . . . . . . . . . . . . . . 2001

*Special Bibliographic Notes related to special journal issues (separates) and indexing/abstracting:*

- indexing/abstracting services in this list will also cover material in any "separate" that is co-published simultaneously with Haworth's special thematic journal issue or DocuSerial. Indexing/abstracting usually covers material at the article/chapter level.
- monographic co-editions are intended for either non-subscribers or libraries which intend to purchase a second copy for their circulating collections.
- monographic co-editions are reported to all jobbers/wholesalers/approval plans. The source journal is listed as the "series" to assist the prevention of duplicate purchasing in the same manner utilized for books-in-series.
- to facilitate user/access services all indexing/abstracting services are encouraged to utilize the co-indexing entry note indicated at the bottom of the first page of each article/chapter/contribution.
- this is intended to assist a library user of any reference tool (whether print, electronic, online, or CD-ROM) to locate the monographic version if the library has purchased this version but not a subscription to the source journal.
- individual articles/chapters in any Haworth publication are also available through the Haworth Document Delivery Service (HDDS).

# A Guide to Docutek, Inc.'s ERes Software: A Way to Manage Electronic Reserves

## CONTENTS

Introduction     1
*James M. McCloskey*

Docutek: Past and Future     5
*Alberta Davis Comer*

ERes: How an Instructional Technology Department Is Only
    as Effective as Its Resources     11
*Win Shih*

Migrating to a New Reserve System: Implementing Docutek's
    ERes System     31
*Madeleine Bombeld*
*Daniel M. Pfohl*

Penfield Library Electronic Reserves Initiative: A Primer
    for Electronic Reserves Service     43
*Andrew Urbanek*

Embracing Fair Use: One University's Epic Journey
    into Copyright Policy     65
*Sandra L. Hudock*
*Gayle L. Abrahamson*

Electronic Reserves, Library Databases and Courseware:
    A Complementary Relationship     75
*Steven J. Bell*
*Michael J. Krasulski*

A Consideration of Docutek's Electronic Reserve System
in a University's Courseware Environment     87
*Donna H. Ziegenfuss*
*James M. McCloskey*

Docutek's ERes Electronic Reserve Software: An Evaluation     99
*Bud Hiller*

Index     119

## ABOUT THE EDITOR

**James M. McCloskey** is Head of Public Services at the Wolfgram Memorial Library, Widener University, Chester, PA. In his current position, he oversees circulation, interlibrary loan, and the reserves operation including electronic reserves. Previously, he managed document delivery services at the Biomedical Library, University of Pennsylvania; managed the Professional Library at the Delaware State Hospital; and served as Information Specialist at BIOSIS in Philadelphia, PA. He has written several articles on the topics of interlibrary loan and electronic reserves. His degrees are from the University of Delaware (BA) and the University of Maryland (MLS). He is currently studying in a distance master's program (MS, Education) through Shenandoah University.

# Introduction

James M. McCloskey

Libraries have, historically, placed high-demand materials on reserve to ensure, at a minimum, short-term availability of these materials to their patrons. Academic libraries in particular have developed reserve systems to ensure availability of materials for students enrolled in specific courses. The time-space limitations of print, however, coupled with technological improvements have led to the use of digitized or "electronic" reserves in many libraries. Docutek, Inc., founded in 1994, has evolved to become a prominent player in the niche market for electronic reserves. Despite the bursting of the Internet bubble in 1999, Docutek emerged from the dot.com bust a resilient and thriving service-oriented company, improving upon their solid electronic reserve system with features such as "Virtual Reference Librarian" and other logical add-ons. This volume is devoted to looking at how several academic libraries selected, implemented and now use Docutek's "ERes" system in order to provide a vital service to their constituency. Alberta Davis Comer provides us with insight to the history of Docutek, Inc. and ERes with an in-depth interview of Docutek founder, Dr. Philip R. Kesten. Her article, entitled "Docutek: Past and Future" reveals how the company was formed as "an opportunity to offer a useful tool to the academic world." Andrew Urbanek, of SUNY Oswego's Penfield Library, describes the "birth pangs" of implementing ERes. Using a "learn by

---

Docutek and ERes are registered trademarks of Docutek, Inc.

[Haworth co-indexing entry note]: "Introduction." McCloskey, James M. Co-published simultaneously in *Journal of Interlibrary Loan, Document Delivery & Electronic Reserve* (The Haworth Information Press, an imprint of The Haworth Press, Inc.) Vol. 15, No. 1, 2004, pp. 1-4; and: *A Guide to Docutek, Inc.'s ERes Software: A Way to Manage Electronic Reserves* (ed: James M. McCloskey) The Haworth Information Press, an imprint of The Haworth Press, Inc., 2004, pp. 1-4. Single or multiple copies of this article are available for a fee from The Haworth Document Delivery Service [1-800-HAWORTH, 9:00 a.m. - 5:00 p.m. (EST). E-mail address: docdelivery@haworthpress.com].

http://www.haworthpress.com/web/JILDD
Digital Object Identifier: 10.1300/J474v15n01_01

doing" pilot project, they resolved problems with reserve indexing, database linking, bandwidth issues, workflow, and copyright. I am sure you will find as you read that they are engaged in a process of continuous quality improvement. At the UNC Wilmington's William Madison Randall Library, Madeleine Bombeld and Daniel M. Pfohl describe their migration "magic" from an ILS electronic course reserve module to Docutek's ERes. They describe how through ERes they streamlined procedures for creating and managing reserve items and courses, improved their copyright management picture, and more efficiently compiled usage statistics. Sandra L. Hudock and Gayle L. Abrahamson of Colorado State University delve into the fair use issues of electronic reserve documents. Walking us through a preliminary review of copyright by various university committees and administrators, this article shows us how the challenges of revising copyright policy are "inversely proportional to ERes' functionality." Lest anyone think that ERes is close to nirvana, Bucknell's Bud Hiller evaluates ERes from a technical perspective. His article considers some of the difficulties embedded in the programming of ERes and "Items that I'd like to change if I ran the world (of Docutek)." Conversely, Win Shih's article considers the merits of the ERes system and walks us through the hardware and network issues involved in setting up ERes on a local campus. Steven J. Bell and Michael J. Krasulski consider the complimentary relationship among ERes, library databases and courseware as do Donna H. Ziegenfuss and James M. McCloskey. Many librarians will find much wisdom contained in these articles and can make informed decisions about the how and why of selecting and implementing an electronic reserve system as well as the academic environment in which such systems operate.

I believe that any library considering or engaged in electronic reserves, whether they use Docutek's ERes or not, will benefit from reading these articles.

Many thanks are extended to Les Morris, editor of the *Journal of Interlibrary Loan, Document Delivery & Electronic Reserve* for his patience and encouragement during this project, and Alberta Comer for sharing her insights and experience.

## *EDITOR'S ADDENDUM*

In June 2004, Docutek announced a significant upgrade to ERes. According to Philip Kestin, Docutek founder, the new release is the culmination of more than two years of software development. It addresses the requests and suggestions made by librarians, faculty, and students, and includes more than 50 of the most requested features by Docutek ERes users over the past three years.

The changes from the previous to the new version of Docutek ERes v5 fall into four categories. Perhaps most obvious to users, the workflow for many common operations is more streamlined. As part of these workflow changes, the layout of on-screen information has been reorganized; for example, Docutek ERes will now use on-screen tabs and information "panels" to provide faster access to documents and data. Users will also notice a significant change in the way copyright-protected documents are managed. The administration of these documents has been moved away from the document level and is based on transactions and course usage in order to correspond more closely with the way libraries deal with copyright. In addition, a host of new features have been added to the high-level administrative aspects of the system. Finally, the underlying database architecture has been upgraded and the software re-engineered to use Visual Basic .NET–widely recognized as the best Web systems technology–in order to improve efficiency.

To improve the workflow, the on-screen layout has been revised to minimize the amount of screen navigation required to complete tasks and find information. Screens will now display a series of either layered or hidden "panels" of information; users can select between panels with a single mouse click. This display philosophy has the intended effect of reducing the number of screens required to complete operations–for example, adding a document and bibliographic information are now completed on either one or two screens.

As part of the workflow and screen layout changes, Docutek has added the capability for libraries to customize nearly every aspect of the Docutek ERes display, including on-screen labels, system colors, and fonts. Many labels in the system, such as "Department," "Course," "Document" and "Account" can now be customized to match the terms used locally by a library. Plus the user can customize many of the administrative views, for example, the ordering and information displayed when searching for copyright-protected documents.

Librarians will notice many differences in the way copyright-protected documents are handled in Docutek ERes v5. For example, full bibliographic information is always directly available to students and others. MARC records can now be automatically generated, and the system can hold an unlimited number of publisher/rights holder letter templates. But perhaps the biggest change is in the philosophy that underlies the management of copyright. Based on requests from customers, the administration of copyright is now based on transactions and course usage. This means, for example, that the status of a document can now be set to "Claim Fair Use" for its use for one course and "Permission Granted with a charge" for another.

A host of new features have been added to Docutek ERes v5, but with careful attention to keeping user interface screens simple and intuitive. For example, administrative actions, such as setting course page expiration dates, have "bulk" options that enable a single operation to act on any number entries at once. A new e-mail interface in Docutek ERes v5 allows faculty and students to get e-mail notifications when changes are made to a course page, for example, when a document is added. And every course page in the system includes an e-mail distribution list, to facilitate electronic communication between faculty and students.

In Docutek ERes v5, the library can set a maximum file size for documents, to prevent large uploads. Also, a new storage allocation report gives a breakdown on the number of courses, documents, and total disk usage (by system, instructor, or department) in order to help librarians better manage the system. There are also upgrades on course pages and with documents themselves. For example, course pages now support an unlimited number of levels of subfolders, to provide more logical organization of documents when necessary. In addition, any number of files can be associated with a single document in the system, something that librarians and faculty have eagerly awaited.

Many of the new features take advantage of the upgrading of the architecture of the database which underlies Docutek ERes, and of the change in software base to Microsoft .NET. These changes enable the user to upload more than one file at time–unlike the previous version of Docutek ERes that had to impose a limit of one file at a time for technical reasons. In addition, the system now supports external SMTP servers, which means one less service running on the local server. In Docutek ERes v5, data from across the system can now be exported into four different file formats, including Excel and tab delimited.

Docutek ERes v5 is the most powerful and most robust version yet of Docutek's standard-setting electronic reserves system. The new release provides a host of new features, but careful attention has been paid to insure that the user interface remains simple and intuitive.

# Docutek:
# Past and Future

Alberta Davis Comer

**SUMMARY.** Dr. Philip R. Kesten, a physics professor at Santa Clara University in Santa Clara, California, co-founded Docutek Information Systems, the parent company for the popular electronic reserves product ERes, in the mid 1990s. This article looks at how and why Dr. Kesten became involved with electronic reserves and where the company hopes to go with its product. *[Article copies available for a fee from The Haworth Document Delivery Service: 1-800-HAWORTH. E-mail address: <docdelivery@haworthpress.com> Website: <http://www.HaworthPress.com> © 2004 by The Haworth Press, Inc. All rights reserved.]*

**KEYWORDS.** Docutek Information Systems, ERes, electronic reserves, Philip R. Kesten

## *INTRODUCTION:*
## *DEFINING ELECTRONIC RESERVES*

Jeff Rosedale (2002) defines electronic reserves thus, "Electronic reserves involves some combination of creating, storing, organizing, pro-

Alberta Davis Comer is Lending Services Librarian, Cunningham Memorial Library, Indiana State University, Terre Haute, IN 47809 (E-mail: libcomer@isugw.indstate.edu).

Docutek and ERes are registered trademarks of Docutek, Inc.

[Haworth co-indexing entry note]: "Docutek: Past and Future." Comer, Alberta Davis. Co-published simultaneously in *Journal of Interlibrary Loan, Document Delivery & Electronic Reserve* (The Haworth Information Press, an imprint of The Haworth Press, Inc.) Vol. 15, No. 1, 2004, pp. 5-9; and: *A Guide to Docutek, Inc.'s ERes Software: A Way to Manage Electronic Reserves* (ed: James M. McCloskey) The Haworth Information Press, an imprint of The Haworth Press, Inc., 2004, pp. 5-9. Single or multiple copies of this article are available for a fee from The Haworth Document Delivery Service [1-800-HAWORTH, 9:00 a.m. - 5:00 p.m. (EST). E-mail address: docdelivery@haworthpress.com].

viding access to, and managing digital objects representing items that faculty have selected to be used directly in conjunction with their instructional activities" (p. vi). Such an electronic reserves (Electronic reserves) system is not a new phenomenon. Today, as many academic libraries who are only now beginning to research the possibilities of electronic reserves discover, one product has established a major presence in the market: Docutek ERes.

## DOCUTEK ERes:
### THE NAME BEHIND THE PRODUCT

Dr. Philip R. Kesten, a physics professor at Santa Clara University in Santa Clara, California, co-founded Docutek Information Systems in the mid 1990s. In a recent telephone interview (October 30, 2003), Dr. Kesten described how a physics professor could become the creator of a library product that has been accepted by parts of the library world.

Kesten first became interested in computers in 1971 when his father was a graduate student at Syracuse. He accompanied his father to the university computer facility and became proficient with the computing process. While his father worked on his dissertation, young Kesten realized that his father dealt with tens of thousands of pieces of information written on 3 by 5 note cards. Kesten wrote the computer code for Card File, a system that his father could use to input these pieces of information and retrieve them as needed. Soon, Syracuse University began using his software. Kesten continued experimenting with computers to find ways to help with school-related projects. He later received a bachelor's degree in physics from Massachusetts Institute of Technology and a doctorate in physics from the University of Michigan. However, his earlier fascination with computers and their potential to help with information management projects continued.

## DOCUTEK'S CONCEPTION

Although Kesten was interested in computers, it is still a leap from being a physics professor to co-founding a very successful electronic reserves system. In 1992, while teaching large classes of physics students at Santa Clara, his students informed him that they needed to connect with other class members outside of class. From these discussions with his students, he created a software package, Phys_Chat. Faculty at

Santa Clara used the software to allow students to participate in electronic study groups. This program became very popular with students and faculty and Kesten eventually shared the software with faculty at more than one hundred other colleges and universities.

Faculty eventually began to query Kesten on other uses for the software. In particular, faculty wanted a way to electronically distribute information to their students. At this point, Kesten said, a light bulb went off in his head. Up until now, he and other faculty had made paper copies of problems, homework assignments, class notes, etc., and put them in the library's reserve room. Students could go to the reserve room, request the items which would be checked out to them, and then they would copy and return the items to the reserve room for other students to use in the same fashion. Students, of course, were thus limited by place and time and could only access the material in the library during library hours. However, Kesten realized, if faculty could take their handwritten material and put it on the Web, students would have access to the material when and where they needed it as long as they had access to the Web.

Kesten asked one of his students, an undergraduate computer science major by the name of Slaven Zivkovic, to help develop a program to use the Web as an electronic reserve repository. Santa Clara University loaned Kesten and Zivkovic a test Web server and they developed the prototype application. In the spring of 1996, the system was available for Santa Clara University faculty to use. Although the system was not advertised except by word-of-mouth, by the end of spring of 1996 over one hundred faculty members were using the system. Kesten and Zivkovic decided that if the system was as good as the faculty and students thought, they should go into business together, not so much to make money but as an opportunity to offer a useful tool to the academic world. Later that year, Kesten and Zivkovic formed their company, Docutek. ERes was their first product and it remains their most popular one.

At this point, many people would have sold the product directly to university faculty, bypassing the library, but Kesten chose to share his product with the library world. Santa Clara University's library was the first to use the ERes system. One of the first universities to actually purchase the system was the University of California at Santa Barbara. There, Kesten and Zivkovic learned one of their first lessons in business when the library director told them they were not charging enough for the system. He said they needed to charge more or else people would not realize the great program they were offering.

When asked what made Docutek succeed when so many companies do not, Kesten replied that it is not that they know some kind of secret; it is just that their approach is different from that of large companies. Kesten said that their technique is to use their company's small size to ensure that they are nimble and light so they can quickly respond to customers' needs. Even more importantly, he said, their motto is to get it right the first time so that problems down the line are small. That way they do not need a large staff of people to fix "bugs." Instead, Docutek staff can devote their time to improving the product.

## PEERING INTO THE CRYSTAL BALL: DOCUTEK'S FUTURE

Although Kesten is spending less time on the business end of Docutek, he still spends time contemplating where technology is going and thus where Docutek needs to go. ERes version 5 was released at the American Library Association's (ALA) midwinter conference in San Diego in January 2004. This new version will incorporate the workflows that users have requested. For example, now library staff must click back and forth between ERes pages to add a document. With the new version, the work can be accomplished from one page. Docutek is also interested in the K-12 market that Kesten says has few solid technology tools. Docutek is putting together an ERes type of tool for elementary and secondary school children and their teachers and parents.

Kesten is also interested in how ERes could be used in conjunction with courseware such as Blackboard. Some libraries are using ERes as courseware for distance education students, but Kesten notes that the trend at many universities is for instruction technology (IT) to use Blackboard (or some other type of courseware), while the library uses ERes. The two software packages, although they could complement each other, are kept totally separate.

## CONCLUSION

Although Kesten modestly bills himself as a non-librarian who has learned much from librarians and the library world, he proudly revealed that he is a card carrying ALA member. In his article "Perspectives of an Enlightened Vendor," he states " . . . it has been both exhilarating and immensely rewarding to witness the explosion in the deployment of electronic reserves systems and to have played a small role in it" (p. 166). As computer technology becomes a more prevalent component of the library world, users can be sure that Kesten and Docutek will continue to play an active and trend-setting role in this brave new world.

## REFERENCES

Kesten, P. (2002). Perspectives of an Enlightened Vendor. In J. Rosedale (Ed.), *Managing Electronic Reserves* (pp. 148-167). Chicago: American Library Association.

Rosedale, J. (Ed.) (2002). *Managing Electronic Reserves*. Chicago: American Library Association.

# ERes:
# How an Instructional Technology Department Is Only as Effective as Its Resources

## Win Shih

**SUMMARY.** The success of a 24x7 electronic reserves service requires a well configured and robustly managed system. Based upon a five-year successful experience of supporting Docutek's ERes system at Saint Louis University, this paper discusses and examines the technical issues affecting the proper operation of the system. Areas covered include selecting an adequate server and scanner; proper server management; reducing PDF file size using various Acrobat 6 features; and interoperability between ERes and other systems. *[Article copies available for a fee from The Haworth Document Delivery Service: 1-800-HAWORTH. E-mail address: <docdelivery@haworthpress.com> Website: <http://www.HaworthPress.com> © 2004 by The Haworth Press, Inc. All rights reserved.]*

**KEYWORDS.** Electronic reserves, ERes, document delivery, interlibrary loan, scanners, scanning, server management, Adobe Acrobat, PDF, Saint Louis University

Win Shih is Assistant University Librarian, Saint Louis University Libraries, 3650 Lindell Boulevard, St. Louis, MO 63108 (E-mail: shihw@slu.edu).
Docutek and ERes are registered trademarks of Docutek, Inc.

[Haworth co-indexing entry note]: "ERes: How an Instructional Technology Department Is Only as Effective as Its Resources." Shih. Win. Co-published simultaneously in *Journal of Interlibrary Loan, Document Delivery & Electronic Reserve* (The Haworth Information Press, an imprint of The Haworth Press, Inc.) Vol. 15, No. 1, 2004, pp. 11-30; and: *A Guide to Docutek, Inc.'s ERes Software: A Way to Manage Electronic Reserves* (ed: James M. McCloskey) The Haworth Information Press, an imprint of The Haworth Press, Inc., 2004, pp. 11-30. Single or multiple copies of this article are available for a fee from The Haworth Document Delivery Service [1-800-HAWORTH, 9:00 a.m. - 5:00 p.m. (EST). E-mail address: docdelivery@haworthpress.com].

http://www.haworthpress.com/web/JILDD
© 2004 by The Haworth Press, Inc. All rights reserved.
Digital Object Identifier: 10.1300/J474v15n01_03

## INTRODUCTION

Information technology has transformed higher education from "just-in-case" delivery of academic resources, to "just-in-time" to "just-for-you," enabling learners to "one-stop shop" for their educational services. In the networked academy, the digital library applies new technology and knowledge management in the collection, organization, production, delivery and preservation of intellectual content directly to the desktop of end users.

As library resources become ever more virtual and ever less physically paper-reliant, our faculty and students are naturally growing accustomed to and demanding to conduct their research online. With this change of the learning and scholarly communication process, the library not only spends more on acquiring online materials, but also allocates more resources to facilitate the access of these materials. Services such as electronic document delivery, electronic reserve, digitizing unique collections, direct full-text linking and remote access authentication are some of the examples. Whether these systems are developed locally or purchased from third-party vendors, they tend to run on a dedicated server (hardware), which is likely managed by the library's instructional technology staff.

The very nature of a digital library is its 24x7 availability and the stability and reliability of its resources. The success of such an operation depends on a variety of factors. A well-configured and managed system is a pre-requisite. A robust and secured instructional technology infrastructure, as well as a competent instructional technology team is also essential. Based upon a successful five-year experience of supporting the ERes system at Saint Louis University, this article shall discuss and delve into the technical issues affecting and influencing the proper operation of the system and by extension, its patrons, administrators and numerous other clients.

## LITERATURE REVIEW

Literature on the ERes system covers disparate areas. Kesten and Zivkovic, inventors of ERes, describe in detail their design philosophy as well as major features of version 3 of the system. Both Lorenzo and Kesten provide a historical overview of the ERes system, as well as future product development plans for the company. Several authors discuss the process of selecting an electronic reserves system at their library, offering comparisons of various commercial and open source systems, including ERes (Kristof and Klingler, Lu, Nackerud).

As for how ERes is used campus wide, DeWeese and McCabe both share success stories from their institutions. However, implementing a new system is by no means merely a technical matter. It also involves changes in workflow, organizational structure, staff skills and the way patrons access resources. Several authors discuss issues related to the implementation of, as well as innovative ways of using the ERes system (Hiller, Kristof, Landes, Walter).

So what do end users think? A survey by Sellen and Hazard shows that ERes is well accepted by students at SUNY, Albany. Some faculty members likewise commented on how ERes enhances their teaching and student learning activities (Birnbaum, Flowers).

Finally, there are several electronic reserve related resources mentioned by many of the authors cited here. The Association of Research Libraries' (ARL) Electronic reserves in Libraries Discussion List is dedicated to the discussion of issues and practices related to the management of electronic reserves. The list archive (https://mx2.arl.org/Lists/ARL-ERESERVE/) should be consulted first for specific electronic reserve questions. The ERes User Group discussion list, set up by Docutek, is dedicated to and participated by ERes customers. Here is where you can seek information specifically related to the ERes system.[1] Jeff Rosedale's Electronic Reserves Clearinghouse (http:// www. mville.edu/Administration/staff/Jeff_Rosedale/index.htm#News) is the most comprehensive portal for electronic reserve resources.

## PHYSICAL SERVER SELECTION

As the command center or "nervous system" of the product, the ERes server performs several roles. First, it operates as a database server, based on Microsoft Access database management system (DBMS) technology. It further functions as a file server storing all of the course documents. In addition, the ERes Server is also a web server hosting the ERes website.

Older versions of ERes run on a Linux/UNIX platform. However, Docutek made a strategic decision to switch to a Microsoft Windows-based platform when ERes version 4.0 was released. At the core of the newer command center is the Microsoft Windows operating system (OS). Although ERes can run on older Windows NT 4.0 systems, it is recommended that administrators migrate to or operate their ERes system on the Windows 2000 OS or Windows 2003 OS server platform for better security, improved manageability, heightened performance and overall strengthened reliability.

Based on the Spring 2004 Customer List provided by Docutek, a survey was conducted to examine ERes operations at each institution. Among the 398 ERes sites, the URLs or IP addresses of 357 sites (90%) were successfully collected. The other 41 sites were not reachable or verifiable due to various reasons, such as the ERes server was behind a firewall or certain kinds of security settings blocked direct access. As indicated in Table 1, over 85 percent of ERes sites are running the Windows 2000 OS. A few sites are running the latest Windows 2003 system, while a small number of customers remain comfortable with the older Linux/UNIX-based version of ERes.

The type of web server used is closely linked to the OS on which the ERes Server is running. It is not surprising, for example, to see that in Table 2, Microsoft Internet Information Services (IIS) 5.0, incorporated within the Windows 2000 OS, is used by the majority of ERes sites, followed by IIS 4.0 (Windows NT 4) and IIS 6.0 (Windows 2003). Apache HTTP software is used by Linux/UNIX-based server systems, while a handful of sites are running Netscape Enterprise Web Server software on their UNIX-based system.

According to Docutek, many of the most common questions raised by new ERes customers center around the hardware specifications of the physical ERes server itself. In particular, customers want to know:

1. How powerful does my ERes server need to be?
2. How much hard drive space do I need?
3. Can I share my ERes server with other applications?

The minimum requirement of physical server (hardware) specifications provided by Docutek (http://www.docutek.com/support/server.html) is a baseline and should be used as reference only. Unless you are converting an older server to host your ERes system, try to purchase a more robust server that includes or can be upgraded in the future to include, additional power and disk space to accommodate the growth and expansion of your electronic reserve operations for at least the next four to five years. Additional capacity furthermore allows you to handle unexpected needs. For example, Saint Louis University's server was capable of hosting a second ERes Server site for our Health Sciences Center Library after the first year's successful operations. In the future, you might want to consider adding Docutek's VRL*plus* component onto the

TABLE 1. OS Used at ERes Sites

| Server Operating System | Sites | Percentage |
|---|---|---|
| Windows 2000 | 307 | 85.99% |
| Windows NT4 | 28 | 7.84% |
| Windows 2003 | 14 | 3.92% |
| Linux | 3 | 0.84% |
| AIX | 2 | 0.56% |
| IRIX | 2 | 0.56% |
| Solaris 8 | 1 | 0.28% |
| **TOTAL** | **357** | **100.00%** |

TABLE 2. Web Servers Used at ERes Sites

| Web Server | Sites | Percentage |
|---|---|---|
| IIS 5 | 307 | 85.99% |
| IIS 4 | 28 | 7.84% |
| IIS 6 | 14 | 3.92% |
| Apache | 6 | 1.68% |
| Netscape Enterprise | 2 | 0.56% |
| **TOTAL** | **357** | **100.00%** |

same physical server, rather than having to buy and set up a second physical server that will cost significantly more.

## Physical Server Specifications

### Processing Power, Speed and Storage Capacity

*Bigger Brains Mean Faster Request Fulfillment.* The microprocessor or central processing unit (CPU) of your server (measured in units called megahertz or gigahertz–Mhz and Ghz respectively–the more of either of these you have, the faster your server) functions as the "brains" of the computer, processing all those electronic reserve requests from your patrons and executing every human command as lightning-fast as humanly possible. Nowadays, forget about megahertz. If you are buying a new server, get as many *gigahertz* as you can afford (1 gigahertz = 1,000 megahertz). "Moore's Law" clearly postulates that the number of transistors per integrated circuit doubles every 18 months. This implies that the processing horsepower or speed of every new computer rolling out of Dell or HP or wherever, doubles every year and a half. Given the seeming accuracy of Moore's Law, a Pentium IV or Xeon processor should drive your ERes Server system and drive it well. Get the best and the fastest machine you can afford.

*Your Computer Can Never Have Too Much Memory.* Yet, at the same time, do not forget this corollary to the mantra, "A computer is only as fast as its slowest part." Random Access Memory or RAM is the working memory which stores data temporarily while your physical server and the CPU are "working on it," running application programs, etc. RAM is to your CPU what a great nurse is to an ace surgeon. The more RAM you have, the more efficiently and expeditiously the CPU runs. The more physical memory you have on the server, the better it will handle multiple requests from users and overall, the quicker your server will "rock and roll." Invest as much in RAM chips (512 MB, 1 GB, 2 GB) as you can afford and that your server will accommodate.

*Decreased Storage Costs Means Good News for Everyone.* The unit price of physical hard drive storage space has dropped dramatically in recent years. According to "Storage Law," optical storage costs decline at the rate of 50 percent every 16 months. Thus, the cost of one gigabyte of raw hard drive space is approaching exactly $1.00 now. Now combine "Storage Law" with "Disk Law." "Disk Law" predicts disk density (the relative "thickness" or "storage capacity" of the magnetic media upon which you house all those jillions and jillions of gigabytes) and increases 25 percent every 12 months, meaning, in a nutshell, that combined with "Storage Law," the hard drive you buy today will store approximately four (4) times the amount of information it could a year ago when you bought it, at nearly half the cost you paid for it one year ago.

As the electronic reserve business burgeons, you want to budget enough room for future growth. Our experience at Saint Louis University indicates that the number of articles scanned into ERes increased more than 350% over a three-year period, with an average of 700 kilobytes stored per article (mostly as Adobe PDF files).

Although the minimum hard drive space recommended for startup by Docutek is 9.1 gigabytes (GB), you will be hard-pressed to find a new server with anything *less* than 18 GB of storage available. That extra hard disk space can easily be allocated for your ERes operations and you do not have to remove old documents as often at the end of each semester in order to make space for new ones. Instead, you can just archive them and suppress them from display to the public.

However, hard drives do crash and they do become corrupted, just by virtue of the fact that they have components which move, produce friction and fail. For obvious safety and redundancy purposes, your mission-critical ERes server should contain at least two hard drives of the same size, with one configured to "mirror" the other one. "Mirror" simply means "shadow" or duplicate. Within a mirrored environment, the data on your primary hard drive is imaged identically onto the second drive. If one drive fails, there will always be a backup working copy of your essential ERes data on the other drive. Budget permitting, it's a wise idea to invest in a fault-tolerant disk system, such as RAID-5 (Redundant Array of Inexpensive Disks–Level 5), where data volumes are "striped" (interleaved or spread out) evenly and intermittently across three or more physical drives, providing even greater protection against data loss. Moreover, combined with an inexpensive, traditional tape-based backup system (to perform overnight replication, for example, of your entire ERes database; these tape(s) can then be stored offsite or in a fireproof vault or both), plus an uninterruptible power supply (UPS) unit providing continuous battery backup power to your ERes server in the event of a significant power outage.

### To Share or Not to Share

Should you allow other applications besides ERes Server to run on your physical server? Depending on what other applications are running on the same machine along with ERes Server, the impact on your ERes operations will vary. Our experience at Saint Louis University has been decidedly mixed. We have not experienced any problems setting up other non-ERes-related websites coded using Microsoft application server script (ASP programming). However, we *have* experienced problems when the website is constructed with Cold Fusion software.

Another complication of operating a shared physical server is simply that when a problem arises, it is difficult to determine which application caused the problem. Since ERes is a major service and if your budget can easily afford the hardware to run it, it is definitely worth your peace of mind and the cost of maintenance to support a dedicated server restricted for ERes operations exclusively.

### Server Management

Like any operations, a properly configured and well-managed server not only performs efficiently, but also reduces the risks of hardware and software malfunctions, let alone malicious intrusions or attacks. To ensure the health of ERes server, the engine for 24x7 operations, Docutek provides its customers with a detailed set of guidelines for proper system administration. Among the most important of these are:

- Maintain good documentation and a logbook of your server. The documentation should contain detailed hardware specifications, major software applications installed, as well as the phone number of technical support of the manufacturer. Make sure to include your server's serial number and date of purchase. Nowadays, vendors such as Dell and Hewlett-Packard let you retrieve product information, as well as download the latest system files based on the product's serial number. Record any major problems or system alerts, as well as actions taken.
- Keep disk space healthy and lean. As your ERes documents increase and the database expands, the original capacious reservoir will quickly fill up. As the disks become crowded, it takes longer and longer for the server to locate and retrieve the information requested by patrons. It also takes longer to back up the system. Docutek recommends regularly performing "disk defragmentation," a process ensuring data on the disk is better organized and thus retrieval of its data is speedier and more efficient.
- Reduce your vulnerabilities. A properly-defined and proactive patch management program and set of security procedures for your network will reduce the vulnerabilities of that network and enhance the stability of all your operations generally, as well as maintain the complete integrity of your data. Poorly configured and badly managed Microsoft Internet Information Services (IIS) servers have been the major targets of many serious worm and hacker attacks. From "Code Red" to "NIMDA," to the "Denial of Service (DOS)" attacks, Microsoft IIS is notorious for being a relatively insecure and vulnerable web server. ERes Server, running Microsoft IIS, should be configured and perform the following tasks on a continual basis:

  - Installing Only the Minimal Setup of Microsoft IIS. When setting up your Windows 2000 server, the IIS are automatically installed as well. The key here is–do not install components such as Front Page Server Extension, Indexing Service or SMTP, if you are not planning to use them. Also, be sure to turn off unnecessary Internet services, such as "anonymous FTP," and the "default Web site."

- Always Applying the Latest OS Patches. Install Microsoft patches and updates when they are made available. In the past, Microsoft issued patches and critical updates irregularly and frequently. In 2003, there were at least 37 Windows 2000 server patches and in 2002, there were more than 60 Windows 2000 patches. This caused a lot of concerns, complaints and often confusion amongst Microsoft's customers. Because of customers' feedback, in October 2003, Microsoft announced a new process of releasing new updates in a more streamlined fashion. Instead of publishing updates irregularly and unpredictably, they are now released on the second Tuesday of every month. These scheduled updates enable server administrators to install multiple updates with a single install and a simple reboot. Of course, if there is immediate risk of security problems, Microsoft will release the necessary patch or update outside of its scheduled timeframe, which is understandable. We have found that the best way to keep informed when there is a new release or update from Microsoft is to subscribe to Microsoft's free e-mail notification service (subscribe at: http://www.microsoft.com/technet/security/bulletin/notify.mspx). You will receive an e-mail notification whenever there is a critical update or a major security alert from Microsoft. This is really the best way to keep your server well patched and secured.
- Continuously Maintaining Proper Protection. Ensure that the anti-virus software on your physical server is configured properly. Make sure that your latest virus definition file is up-to-date. All major anti-virus software allows you to configure your system automatically to download and install the latest virus definition files, as well as to scan your system at scheduled intervals. Take advantage of all of these protective, timesaving mechanisms.
- Maintaining Proper Physical Server Configuration. Close unnecessary network ports. Information and requests are communicated through certain ports. For example, web servers typically "talk" on port 80. Be sure to close ports that are not used by ERes and other applications running on your server. This will reduce the possibility of being attacked through unused ports. Change the default home directory of

your ERes site from "c:\inetpub\wwwroot" to a different path and hard drive partition.
- Subscribing to the ERes User Group Discussion List. From this list, you will receive helpful alerts, product updates and news from Docutek and fellow colleagues regarding any ERes related topics.

## SCANNERS AND SCANNING SOFTWARE

### Selecting the Scanner That's Perfect for You

Scanner price varies from a low-end, consumer model costing under $100, to high-speed, commercial-grade behemoths ranging over $10,000. Staff must choose the appropriate scanner with "futurethink" in mind. Choose a scanner that will meet your future scanning quantities and, within budget too, can be a daunting decision.

### Major Selection Criteria Include

### Scanning Speed

When you have a large number of reserves to be scanned at the beginning of a semester, speed makes a major difference. Our Saint Louis University stats clearly demonstrate that the average reserve article is 14 pages. For a 14-page article, it takes 2.8 minutes to scan on a five-page-per-minute (ppm) scanner, compared to about 34 seconds on a 25-ppm scanner. The difference is 2.24 minutes. If your library scans 500 articles per semester with all else being equal, you will save a total of 18 hours (1,120 minutes) of scanning time using as fast a scanner as possible. It is important to think about the time you save as well as the staff salary dollars conserved (see Table 3).

### Scanning Resolution

A scanner's resolution or reproduction quality is measured in dots per inch (dpi). The higher the dpi, the better the resultant image quality. Because higher resolution quality connotes more information (more dots), the size of the file created will be larger given the greater dpi. By extension, it will take patrons longer to download larger files from your ERes website. For example, when scanning at 300 pixels per inch a

TABLE 3. Scanning Speed Comparison

| Scanning Speed | Time Required for Scanning a 14-Page Document |
| --- | --- |
| 5 pages / minute | 2.8 minutes |
| 10 pages / minute | 1.4 minutes |
| 25 pages / minute | 0.56 minutes (34 seconds) |

page of 6 inches across by 8 inches down, the total pixels scanned will be the product of horizontal pixels (300 × 6) vertical pixels (300 × 8), which is 4,320,000 or 4.32 million pixels. The same page scanned at 100 dpi, the total pixel count drops to 480,000 or .48 million pixels, one-ninth the size of the 300 dpi scan. Thus, file size is proportional to the square of the image resolution. If an article to be scanned is text-based or if the quality of graphics, charts and other pictorial information within the article are not critical in terms of their "high legibility," it is recommended by Docutek to scan your articles at 75 to 150 dpi.

Clearly, file size of your documents can be a determining factor in the overall satisfaction of your customers and the ultimate success of your ERes operations.

## The ("Must Have") Automatic Document Feeder (ADF)

This an indispensable feature when scanning vast quantities of pages and is highly recommended as a "must have" add-on feature by Docutek. Instead of loading page after page after page of article materials manually, the automatic document feeder empowers you to scan multiple page documents nonstop. Your typical ADF will accommodate up to 50 through 100 pages, enabling you to go on and do other things without wasting valuable time and energy feeding source pages to the scanner.

## Paper Size

Most flatbed scanners are designed to handle letter-sized articles (8 1/2" × 11"). Some can go beyond this and handle legal-sized documents (8 1/2" × 14"). Scanners with larger flatbeds obviously can process larger-sized documents, but be prepared to pay more.

*Duplex Capability*

Instead of scanning one side of a document at a time (simplex mode), the ability to scan both sides of a page simultaneously (duplex mode) will save you and your staff further time and energy, especially in dealing with two-sided documents. Document scanners that can scan in duplex mode are frequently described in the product specifications as "ipm (images per minute)." Therefore, 20 ppm on a simplex scanner may be rated as 40 ipm on a duplex scanner. As expected, duplex scanners will cost you more, but the price offset may be well worth the pain to your pocketbook if you frequently deal with double-sided documents.

*Color Scanning*

The capacity for rendering color may be quite crucial for some documents such as works for an art history class, pie charts for a course in business statistics, slides of the human body or cells for a medical seminar. At Saint Louis University's Health Sciences Center Library for example, we operate a Fujitsu 4210C Color Scanner with ADF that can scan color documents at a rate of 25 ppm, with a resolution of 150 dpi.

### Selecting the Best Scanning Software

*To Capture or Not to Capture*

Besides a decent scanner, you need the Adobe Acrobat software application to convert paper images into PDF files. Some libraries opt to use another Adobe product, called Adobe Capture, to create Adobe files. Adobe Capture is designed to convert a higher volume of paper documents into searchable Adobe PDF files. It also offers OCR (Optical Character Recognition) capability, automatic page and content recognition and powerful clean-up tools. However, version 6 of Adobe Acrobat, which released in 2003, also includes a built-in OCR feature, called Paper Capture. At Saint Louis University, we use Adobe Acrobat for our ERes scanning operation.

*Size Does Matter*

Since the file size of PDF documents directly impacts the speed at which patrons can download and display these documents to their desktop, the smaller the file size, the less time your patrons have to wait. Fortunately, Version 6 of Adobe Acrobat introduces a series of new and improved ways, as described below, to optimize the file size of PDF files.

*Use the "Save As" Command.* For PDF files created with older versions of Acrobat, use the "Save As" option from the "File" menu to save your document with the same name. Acrobat will overwrite the original document in a more efficient way. Furthermore, it will allow the file to be downloaded one page at a time from the web, by virtue of a new feature called "Fast Web View." In Acrobat 6, that feature is enabled by default and permits the document to appear more quickly at the user's end instead of displaying only after the entire document has been downloaded.

*Compatibility Level.* The easiest way to ensure compatibility of your documents is to use the "Reduce File Size" option from the "File" menu when saving your documents. You can set the file compatibility level from Acrobat version 4.0 to 6.0. If you are sure that none of your users will be using versions of Adobe Reader older than 4.0, this will be a good way to keep file sizes as small as possible.

*PDF Optimizer.* This new feature is only available in Acrobat 6 Professional, not the standard edition. PDF Optimizer improves the performance of PDF files by allowing you to modify various Adobe settings, including compression methods, fonts embedded in the document and bookmarks, forms, comments and thus effectively reduces file sizes.

*Paper Capture.* All paper documents are initially scanned into Acrobat as bitmap images. Even though they are displayed as text through Acrobat Reader, they are just pictures of text and are not searchable or editable. The new "Paper Capture" feature of Acrobat 6 not only can convert scanned image files into searchable text files, but also potentially reduce PDF file sizes as Adobe Capture does. Its built-in dictionaries support 16 languages. Adobe 6 offers three PDF output styles when performing OCR.

1. "Searchable Image Exact" will make the image identical to the original image (foreground) and place the searchable text behind the image. Thus, the file size here tends to be larger than the other two options. It will not allow users to edit or modify existing text. Yet this option is ideal for documents such as contracts and legal documents whose integrity you wish to maintain, but also offer search capability of its contents.

2. "Searchable Image Compact" lets you further reduce the scanned image file size through a compression technique, such as JPEG for color images and ZIP for black and white, to the foreground image. Choose this option when both file size limitation and search ability are important.

3. "Formatted Text and Graphics" will substitute the scanned image with searchable, editable vector characters, similar to the process of PDFMaker (a macro for converting Microsoft Office documents to PDF files) or Acrobat Distiller (a standalone application that creates PDF files by simulating a printer). If the OCR engine cannot recognize a word or the word is not in its dictionary, it will remain as a bitmap and is not converted to vector-based text. The file size will consistently be the *smallest* among the three options listed here. However, since the original image is replaced by "OCRed" text, mistakes are unavoidable. "Paper Capture" provides a utility for you to go over all suspicious words and correct them as necessary. However, the interface is clumsy by comparison to "Adobe Capture."

To better understand each of the file size reduction options mentioned above, an experiment was conducted. Ten PDF documents originally produced by Adobe Acrobat 4 were randomly selected. They were retrieved into Adobe 6 and then processed and saved with each of the options mentioned above. The results, shown in Table 4, provide a good comparison/contrast of these file-size reducing techniques.

As Table 4 indicates, the average size of a document reduced in size varies from a little more than 8 percent to close to 70 percent. If search ability is not a mandatory factor and all of your users are equipped with at least Adobe version 4 or higher, method B probably is the optimal option. If searching text is obligatory and you are willing to invest the time and effort to correct OCR "suspects," then the three Paper Capture options (C thru E) are good enough choices. You might want to experiment with all the options in any event and compare the quality of each, the times required and the consistency of results before committing to one method only.

## INTEROPERABILITY

With more campus services and resources available electronically, how to make all these disparate and heterogeneous systems residing on

TABLE 4. File Size and File Saving Techniques

| Document | Pages | Adobe 4 | A | B | C | D | E |
|---|---|---|---|---|---|---|---|
| 1 | 9 | 1232 | 919 | 533 | 493 | 342 | 301 |
| 2 | 9 | 798 | 800 | 465 | 431 | 241 | 225 |
| 3 | 6 | 331 | 333 | 230 | 198 | 94 | 67 |
| 4 | 1 | 104 | 107 | 64 | 107 | 95 | 80 |
| 5 | 14 | 1419 | 1421 | 1045 | 833 | 536 | 560 |
| 6 | 7 | 800 | 691 | 397 | 429 | 206 | 249 |
| 7 | 6 | 638 | 640 | 389 | 334 | 174 | 152 |
| 8 | 9 | 1380 | 1111 | 680 | 587 | 351 | 279 |
| 9 | 9 | 997 | 999 | 584 | 512 | 314 | 241 |
| 10 | 5 | 420 | 422 | 273 | 302 | 249 | 343 |
| Average | 7.5 | 108.25 | 99.24 | 62.13 | 56.35 | 34.69 | 33.29 |
| % Change | | 0.00% | 8.33% | 42.60% | 47.95% | 67.95% | 69.24% |

Adobe 4–File sizes (in Kilobytes) for documents scanned and converted to PDF files using Adobe Acrobat version 4.

Experimental Methods (results in Kilobytes):

A–Saved file as "Acrobat 6 PDF File."

B–Saved file using "Compatibility with Acrobat 4.0 and Later Features" option.

C–Saved file using "Paper Capture with Searchable Image Exact" option.

D–Saved file using "Paper Capture with Searchable Image Compact" option.

E–Saved file using "Paper Capture with Formatted Text and Graphics" option.

multiple domains, running on atypical system platforms, with irreconcilable and incompatible metadata formats, managed by different system owners and requiring separate authentication, work seamlessly is a monumental undertaking. However, ERes can serve as a gateway to many course-related external resources to which libraries subscribe or generate. For the same reason, with proper configuration, ERes also permits other external systems to point to its resources at the course or article level. With such flexibility, libraries can decide the most efficient way to integrate ERes with its other resources based upon their local situation and needs.

## *Integrating ERes with Your Library's Online Public Access Catalog (OPAC)*

Before electronic reserve, information on course reserve collections was kept in a library's OPAC. Patrons consulted their OPAC to find the information and availability of reserve materials for their courses. With implementation of ERes, containing a subset of course reserves, libraries are facing a decision as well as dilemma. What is the best way to integrate the two independent systems? In many cases, the OPAC is still the place to look for traditional reserve items like books, multimedia reserves (audio/video tapes, computer software) and other objects. We actually have human bones on reserve in our Health Sciences Center Library. In a survey of ARL libraries, the capability of integrating with OPAC is one of the desired functionalities of an electronic reserve system (Kristof). Other authors echo this as well (Driscoll, Kristof and Klingler, Laskowski, Lu).

The 300-plus functioning ERes sites provide us with a glimpse of how libraries are dealing with this dilemma. As Table 5 shows, more than 80 percent of the libraries have a direct link from their main home page to ERes' main page. An additional 11 percent of the libraries provide direct links to ERes, but require authentication. In such an arrangement, a library has two options in dealing with reserves:

1.  Maintain a dual system: Keep both ERes and OPAC as separate and independent. Patrons use ERes to find reserves in electronic format and consult the OPAC for traditional reserve materials housed in the library. This is similar to the situation of electronic journals, where some libraries maintain a separate A to Z list or searchable database for electronic journals. However, patrons still go to the OPAC for their printed journal collection. Maintaining a dual reserve system requires less work for the library since there is no need to create any records or links between the two systems. However, a dual system will cause confusion for students. Some libraries do provide instruction on the difference between the two systems on their Reserve website.
2.  ERes functions as the single access point for both electronic and traditional reserve materials. There are links from ERes course pages to item records in the OPAC for traditional reserves. At Saint Louis University, we are in the process of implementing this option.

A minority of libraries, as Table 5 illustrates, prefer that the OPAC remain the single system for all reserve materials. There is no direct link from the library's home page to ERes' main site. Instead, patrons will find and link to ERes items through OPAC. Some libraries, for copyright reasons, further require authentication before patrons can reach the

TABLE 5. ERes and Library OPAC

| Accessibility | Sites | Percentage |
|---|---|---|
| Link to ERes from library's home page with no restrictions | 259 | 83.01 |
| Link to ERes from library's home page with a logon requirement | 36 | 11.54 |
| Access to ERes through library's OPAC. No direct link to the ERes site. | 10 | 3.21 |
| Access to ERes through OPAC with logon requirement. No direct link to ERes site. | 7 | 2.24 |
| **TOTAL** | **312** | **100.00** |

ERes course item. With ERes' MARC Writer utility, a library can generate bibliographic and item records containing URLs of specific ERes items in any of its 856 fields. These records can then be uploaded into the OPAC database.

Supporting a single point of access to reserve materials, either through ERes or the OPAC, requires additional staff time and effort to generate hyperlinks and maintain records on both systems. However, from the vantage point of patrons, it is simpler and more straightforward to locate all reserve items, no matter what their formats, through a single system.

### *Integration with Other Library Electronic Resources*

Several authors discuss the desire of linking their library's electronic reserves system to other commercial resources to which their library subscribes (Armstrong, Cody, Kesten, Smith, Sylvester). With increasing chunks of library resources already earmarked for acquiring, processing and facilitating access to licensed electronic resources, how to efficiently utilize these resources and to link them effectively with existing library systems, for ease of access and interoperability, is "the next big thing" on the table. In his article on their operation of "deeplinking" to online resources from electronic reserves, Warren points out various incentives for linking to library paid subscriptions to full text articles:

- Saving copyright costs. This is especially true considering the recent hike of copyright fees in general.
- Efficient use of subscribed resources for which the library has already paid.

- Reduce duplication of effort. With full text already available online, there is no need for a library to redigitize its articles repeatedly.
- Higher quality of images. The library does not have to rely upon a downgraded photocopied version of its material for scanning.
- Accessibility to patrons with disabilities.

Nevertheless, there are innumerable complications related to each of the above goals and operations. For example, the identification of the structure a vendor used for their URLs, variation of reliability and stability across URLs amongst vendors, basic quality control and staff training are just a few of the ongoing hurdles to overcome. Additionally, compliance with contract agreements between commercial vendors is necessary to configure ERes at certain levels to enable authentication for patrons from non-institutional IP addresses. The placement and location of authentication varies. Some libraries place it at the reserve main entrance level (Nackerud, Walter), while at others it dwells at the level of each article itself (Armstrong, Cody, Laskowski, Sylvester, Warren).

### *Interpenetration with Courseware Systems*

In higher education, traditional classroom instruction is now complemented and supplemented by online course management systems such as Blackboard and WebCT. ERes Server can be configured to establish a "one-way trust," so that the traffic coming from Blackboard or WebCT does not require the regular ERes logon prerequisite. At SLU, we support both Blackboard and WebCT. Therefore, our ERes Server is configured to accept links coming from both courseware servers. Some of our faculty members use one or both courseware systems in conjunction with ERes. They leave all the work related to library reserves, such as scanning, copyright clearance and so forth, to the library. Our reserve staff provides faculty with ERes URLs to include in their course pages on Blackboard or WebCT systems. Their students access ERes reserve materials directly from the links posted on their Blackboard or WebCT course sites. It all works like a charm.

### CONCLUSION

The success of the ERes operation requires a dedicated teamwork approach committed to making it work across disparate functional units and to the satisfaction of a diverse end user base. When we implement a new system, such as ERes, we not only need to ensure that we have adequate hardware, infrastructure and instructional technology expertise, but we also need to adjust our organizational structure, workflow and staff skill sets responsively and responsibly.

From Instructional Technology's point of view, minimizing downtime and maximizing our system resources required tight server configuration, highly effective patch management practices, tamper proof security settings, scheduled backups, regular file archiving, constant disk cleanup and decent documentation. Producing the best quality output possible requires well-dressed scanning equipment and right-thinking software applications. Instructional technology personnel need to work cordially with reserve staff to identify the most efficient workflow and to resolve any technical concerns. Open communications is more than just a buzzword and free exchange of information facilitates ERes operations efficiently so that further innovations are constantly encouraged.

## NOTE

1. To join the Docutek Users Group, simply send an e-mail to: usergroup@ docutek.com with the word "Subscribe" in the subject field. Unfortunately, because of our Web site reorganization, if you had already subscribed to the Users Group you will need to subscribe again.

## BIBLIOGRAPHY

Armstrong, William W. and Peggy P. Chalaron. "The electronic reserves program at LSU: A dual study in service and survival." *Journal of Interlibrary Loan, Document Delivery & Information Supply* 11 No. 4 (2001): 1-23.

Birnbaum, Matthew. "First response: The electronic reserve experience." *Polk Library News* No. 14 (August 2001) [Online]. Available: http://www.uwosh.edu/library/news/2001/aug01.html.

Cody, Sue Ann, Dan Pfohl and Sharon Bittner. "Establishing and refining electronic course reserves: A case study of a continuous process." *Journal of Interlibrary Loan, Document Delivery & Information Supply* 11 No. 3 (2001): 11-37.

DeWeese, June. "Why did we choose ERes?" Presented at the Second Annual MOBIUS Users Conference, June 1, 2001.

Driscoll, Lori. "Chapter 2: Getting started." *Journal of Interlibrary Loan, Document Delivery & Information Supply* 14 No. 1 (2003): 7-25.

Flowers, Lamont A. "Using Docutek ERes in a student affairs classroom." Student Affairs Online 5 No. 1 (Winter 2004). [Online]. Available: http://www.studentaffairs. com/ejournal/Winter_2004/UsingDocutekERes.htm.

Hiller, Bud and Tammy Bunn Hiller. "Electronic reserves and success: Where do you stop?" *Journal of Interlibrary Loan, Document Delivery & Information Supply* 10 No. 2: 61-75.

Kesten, Philip R. "Perspectives of an enlightened vendor." In: Jeff Rosedale, ed. Managing electronic reserves. Chicago: American Library Association, 2002. pp. 148-167.

Kesten, Philip R. and Slaven M. Zivkovic. "ERes–Electronic reserves on the World Wide Web." *Journal of Interlibrary Loan, Document Delivery & Information Supply* 7 No. 4 (1977): 37-47.

Kristof, Cindy. *Electronic reserve operations in ARL libraries: A SPEC kit.* (May 1999) SPEC Kit #245. Washington, D.C.: Association of Research Libraries, Office of Leadership and Management Services.

Kristof, Cindy and Tom Klingler. "Kent State University's electronic reserves experience." *Journal of Interlibrary Loan, Document Delivery & Information Supply* 11 No. 1 (2000): 39-49.

Landes, Sonja. "Electronic reserves at Milne Library SUNY Geneseo." *Journal of Interlibrary Loan, Document Delivery & Information Supply* 12 No. 1 (2001): 27-33.

Laskowski, Mary S. and David Ward. "Creation and management of a home-grown electronic reserves system at an academic library: Results of a pilot project." *The Journal of Academic Librarianship* 27 No. 5 (2001): 361-371.

Lorenzo, George. "Docutek reflective of growth of academic librarians' changing needs in a digital age." *Educational Pathways* (November 15, 2003). [Online]. Available: http://www.edpath.com/docutek.htm.

Lu, Songqian. "A model for choosing an electronic reserves system: A pre-implementation study at the Library of Long Island University's Brooklyn Campus." *Journal of Interlibrary Loan, Document Delivery & Information Supply* 12 No. 2 (2001): 25-43.

McCabe, James. "The rise of ERes." Inside Fordham Libraries 16 (Fall 2000). [Online]. Available: http://www.docutek.com/products/eres/fordhampub.html.

Nackerud, Shane. "Electronic reserves: Home grown vs. turnkey." In *Proceedings of the 9th National ACRL Conference*, edited by Hugh Thompson. Chicago: ALA, 1999. [Online]. Available: http://www.ala.org/ala/acrl/acrlevents/nackerud99.pdf.

Sellen, Mary and Brenda Hazard. "User assessment of electronic reserves and implications for digital libraries." *Journal of Interlibrary Loan, Document Delivery & Information Supply* 12 No. 1 (2001): 73-83.

Walter, Don. "Using Docutek ERes with EZProxy." Presented at SouthEast Voyager User Group Meeting, July 23, 2002. [Online]. Available: http://www.jsu.edu/depart/ library/personal/walter/Electronic reserves.html.

Warren, Scott. "Deeplinking and Electronic reserves: A new generation." *Journal of Interlibrary Loan, Document Delivery & Information Supply* 14 No. 2 (2003): 65-81.

# Migrating to a New Reserve System: Implementing Docutek's ERes System

Madeleine Bombeld
Daniel M. Pfohl

**SUMMARY.** Randall Library, UNCW, recently migrated from its established electronic course reserve system to Docutek's ERes course reserve system. This article reviews the migration path and discusses decision points and events along the way, the impact on reserve processes and workflow and challenges yet to be met. *[Article copies available for a fee from The Haworth Document Delivery Service: 1-800-HAWORTH. E-mail address: <docdelivery@haworthpress.com> Website: <http://www.HaworthPress.com> © 2004 by The Haworth Press, Inc. All rights reserved.]*

**KEYWORDS.** Electronic course reserves, digital course reserves, electronic reserves, digital reserves, ERes

---

Madeleine Bombeld (E-mail: bombeldm@uncw.edu) is Assistant University Librarian, Coordinating Access & Delivery Services and Daniel M. Pfohl (E-mail: pfohld@uncw.edu) is Associate University Librarian for Computing Services, both at William Madison Randall Library, The University of North Carolina at Wilmington, 601 South College Road, Wilmington, NC 28403-5616.

Docutek and ERes are registered trademarks of Docutek, Inc.

[Haworth co-indexing entry note]: "Migrating to a New Reserve System: Implementing Docutek's ERes System." Bombeld, Madeleine, and Daniel M. Pfohl. Co-published simultaneously in *Journal of Interlibrary Loan, Document Delivery & Electronic Reserve* (The Haworth Information Press, an imprint of The Haworth Press, Inc.) Vol. 15, No. 1, 2004, pp. 31-41; and: *A Guide to Docutek, Inc.'s ERes Software: A Way to Manage Electronic Reserves* (ed: James M. McCloskey) The Haworth Information Press, an imprint of The Haworth Press, Inc., 2004, pp. 31-41. Single or multiple copies of this article are available for a fee from The Haworth Document Delivery Service [1-800-HAWORTH, 9:00 a.m. - 5:00 p.m. (EST). E-mail address: docdelivery@haworthpress.com].

http://www.haworthpress.com/web/JILDD
© 2004 by The Haworth Press, Inc. All rights reserved.
Digital Object Identifier: 10.1300/J474v15n01_04

## BACKGROUND

The University of North Carolina at Wilmington, Randall Library was not a newcomer to the realm of electronic reserve when Docutek's ERes caught our attention (Cody, Pfohl and Bittner, 2001). We had been using the course reserve module of our integrated library system for a number of years before jumping into the electronic reserve business in 1998. Integrating electronic reserve services into the integrated library system was just another step in the ongoing process of making specialized course information available to our patrons.

While the ILS's electronic course reserve module did integrate electronic reserve into the catalog, we continued to have a number of problems with it. The electronic reserve module was cumbersome, requiring bibliographic/item creation in the cataloging module and course creation/linking in the reserve module. Another Windows GUI cataloging and image linking piece of software was required for the processing of the electronic reserve images. This software included proprietary scanning and an image reading applet.

Maintenance of courses was also cumbersome. To remove reserve items and courses required a reverse process through the different modules.

In addition, the integrated library system course reserve module was based on a single reserve database for our three-library consortium. The online catalog displayed reserves for all three libraries as if they were one unit. The ILS's course reserve module left copyright tracking and compliance entirely up to the library. A complicated external process of copyright tracking and compliance was developed. Finally, the statistics compiling and reporting mechanism of the integrated library system reserve module also fell short of what we needed. It required external manipulation of data to isolate basic use counts for each of the libraries.

What we were looking for when we began reviewing Docutek's ERes was a system that would integrate all of our reserve material (traditional and electronic), streamline procedures for creating and managing reserve courses and items, effectively manage copyright compliance and compile and report statistics in an efficient and meaningful manner.

After reviewing the online information about ERes, we requested an onsite demonstration. We were impressed with what we saw. Building reserve courses seemed quick and easy. No cumbersome bibliographic and item files to create first. Any kind of digital document can be added to a course. We would be able to scan documents directly into Adobe

Acrobat to produce PDF files that can be easily transported to the server and attached to courses. Documents could also be faxed directly to the server, automatically saved as PDF files and attached to courses. Adding a link to an Internet site or online catalog entry is as easy as adding the URL to the course. Maintenance of courses and reserve items appeared simple and quick. Copyright tracking and compliance is integrated at the document level and includes a direct link to the Copyright Clearance Center. Reporting also seemed straightforward and would easily provide useful statistics without having to use third-party software to manipulate the data. The user interface (an important aspect for any reserve system) appeared clear, uncluttered and easy to navigate.

The Docutek sales person set up a trial account on an ERes server for our librarians to experiment with the software. The decision to migrate to ERes was made soon after. The proposal to implement ERes was made in December 2002. The purchase orders for the server equipment and ERes software were submitted at the end of March 2003. The server arrived near the end of April. Docutek loaded the ERes software in mid-May. Initial course set up and staff training was completed in time for our proposed deadline of May 30. ERes was scheduled to be operational as the reserve system by August 15, the start of the fall semester.

## *HARDWARE/SOFTWARE CONFIGURATION*

Because we were already using an electronic reserve system, we had a solid starting point for estimating storage space needs and the level of use. Profiling the server needed for ERes was then a matter of coordinating current storage and use statistics with Docutek's recommended specifications and estimating future reserve growth. The University of North Carolina at Wilmington has an enrollment of approximately 11,000 students and a full- and part-time faculty of about 800. Our previous electronic reserve module was serving 150 courses taught by 125 faculty members. PDF storage for this module was about 4 GB. Our long-range plans included converting our paper files (2,500-3,000 items) to electronic format. The statistics we were able to compile showed an average of 4,700 electronic reserve transactions per month.

Docutek's support site has good documentation on server requirements. Docutek's Install Team was very helpful and quick in reviewing UNCW's enrollment and current reserve information and recommending a server profile. We varied from their specification in two characteristics; a larger monitor and 36 GB hard drives instead of 18 GB drives. The larger drives did not add much to the server cost and will ensure ample storage for the effective life of the server. Other specifications include: 1.8 GHz processor, 1 GB RAM, 2 36 GB hard drives in a RAID 1 array, Internal network adapter, Internal modem, DDS4 tape backup, 20/40 GB capacity, and Windows 2000 Server operating system.

Our computer vendor also reviewed the proposed configuration to ensure all selected hardware was compatible. Once satisfied that the server specifications would meet our needs, the equipment was ordered.

Well before the new server arrived, the Docutek Install Team contacted us to begin profiling the software and to set an installation date. They provided specific questions about contacts at the library and server connection information. We provided the installation team with customization information such as how we wanted the library's name to appear, e-mail address, copyright disclaimer, item descriptors for hardcopy items and a list of departments.

Server set up, configuration and maintenance would be the responsibility of the library's Systems Office. The server purchase specified installed RAID configuration and installed Windows 2000 Server operating system. So once the server arrived, set up was fairly easy and included registration on the campus domain, acquisition of a static IP address, implementation of security measures such as running IIS Lock Down Tool, IIS 5 Baseline Security Checklist, Windows 2000 Server Baseline Security Checklist and running the Microsoft Baseline Security Analyzer.

With the server ready to receive the ERes software, several more steps were taken to allow access by Docutek's Install Team. These steps are laid out and explained in Docutek's documentation. The directions describe the required administrative account to be set up in the local computer domain. Remote control software also was installed. Of the several products suggested, we chose Microsoft's NetMeeting, which is part of Windows 2000 Server and is the recommended software. Docutek's technical support team uses NetMeeting to perform the ERes software installation, set up and future troubleshooting. The final step in setting up access for ERes is ensuring access through a firewall. A brief list of TCP ports must be opened for Docutek staff to access the server and for patrons to access ERes. These ports include the web server, FTP, mail server and chat server.

One useful feature of ERes, that requires further discussion, is the ability to fax documents to the server. This, of course, requires a phone line to connect to the server's modem. A server management consideration of this feature is that it requires an administrator level account to be logged on at all times. The best way to implement this securely is to create an administrative account in the local computer domain, set a password for the screen saver and set a brief delay time for the screen saver to start. As faxes are received, they are routed through the Adobe Acrobat Distiller program and posted to the user's staging directory as a PDF file. This ability to fax documents directly to ERes has greatly facilitated the conversion of our paper files to digital format.

We felt that restricting access to courses/documents in ERes using the password feature would be too cumbersome to manage and would result in numerous calls about unknown or forgotten passwords. We desired to make reserves as easy as possible for our patrons, but, because of copyright considerations, we still wanted to restrict ERes access to UNCW students and faculty. We, therefore, employed a proxy rewrite mechanism via our ILS. Thus, anyone accessing ERes from off-campus must authenticate against the Library's patron database.

Once ERes was installed, we needed to make decisions about how to set it up. This meant making decisions on what formats, if any, to use for course descriptions, course numbers and additional information about the courses. We also needed to decide how to enter information about the "account holders," those individuals (faculty) who need accounts in ERes in order to be linked to a particular course. We decided that completing a minimal number of fields on the "Add an Account" screen would suffice and we completed the account level, username, first name, middle initial, last name and e-mail address fields. Only the username, first and last name are required fields.

We examined a variety of options for setting up faculty members' accounts, including setting up accounts for all faculty, for all faculty who had used reserve in the past and/or for all faculty who had courses the last academic semester. The decision was made to identify faculty members who were regular reserve users and set up accounts for those first. Then we added accounts for faculty members as they asked for course reserves. Once an account was set up, the course site could be created but we did not set up any course web sites until faculty members requested course reserves for specific courses.

Another factor we discussed regarding accounts in ERes was whether or not to enable course site management privileges to account holders, i.e., our faculty. ERes is very powerful and allows account holders at certain levels to make changes to course sites. We felt it was important to ensure consistency of data input and a uniform look to course sites and that meant restricting access to reserve staff members. In light of that, we decided that our faculty members account level would be that of ERes' designated Faculty but that we would not train faculty to modify their own course sites or add new course sites. Furthermore, we did not give our faculty their username and password so access to their course sites was restricted as well.

Once the decision was made regarding our faculty account set up, we entered some test accounts so we could walk through the various steps and options offered by ERes. Reserve staff members and the Access Services and Systems Librarians all had accounts set up as managers so they could perform the full range of ERes features allowed by that status. Each person set up courses, created documents, linked documents, modified courses and documents and walked through other potential steps as often as was necessary to gain a solid familiarity with ERes. We were pleased to discover that ERes was intuitive and easy to use and therefore allowed us to move from test environment to live implementation more rapidly than we thought possible.

## NOTIFYING FACULTY OF THE NEW RESERVE SYSTEM

The next challenge was deciding how to notify faculty that we had a new reserve system. We not only needed to announce our new course reserve system but we also needed to communicate that previous reserve lists would not automatically be converted. As most library staff realize, faculty members operate on a timetable that does not necessarily reflect the needs and desires of the reserve staff to submit requests for course reserve early, preferably before the start of the semester. This reality was a concern as we realized that handling course reserve requests in a new system would undoubtedly cause some delays in getting materials on reserve. We expected this to happen despite the fact that ERes greatly simplified the overall reserve process, since we were working with new routines and procedures.

Starting in July we notified faculty members to think about their course reserve requests and to submit those lists as early as possible. This notice appeared in each of our weekly campus newsletters sent to all faculty and staff. Toward the end of July, we posted a news bulletin on the library homepage that described our new course reserve system and asked faculty to submit their course reserve requests. Throughout this period, the old course reserve system was still available on the library home page as we had students who still needed materials identified through that system. We made the switch to ERes on a pre-announced date and once again issued a plea to faculty to submit their course reserve requests. An additional message to faculty and staff went out via campus email and announced the ERes link was live. We once again asked faculty to submit their course reserve requests.

## ESTABLISHING PROCEDURES FOR PLACING MATERIALS ON RESERVE

Early, we decided that all documents added to course reserve would be in PDF format. Since part of our ERes software purchase included the DocuFax module, converting files to PDF format simply required faxing the document to our ERes server. This meant we could accept documents in any format from our faculty. Using the DocuFax module also allowed us to handle a large number of documents in a much smaller amount of time than if we had to scan each one. This was particularly helpful to us as we had a large number of documents stored in files that we needed to convert to PDF format.

There was also an extensive collection of course reserve PDF files already stored on computers and part of our initial discussion of establishing procedures dealt with how to handle requests for those files. We did not want to touch those files unless we had specific course reserve requests that called for their use, but we knew that faculty would request some of these documents. We planned for this by developing a procedure that utilizes one of the more powerful features in ERes, linking to file based documents. Each staff member who worked in ERes had a drive on their computer mapped to the computer that stored these documents. When a request for one of these documents came in, we could browse to that drive, select the document and transfer that file to ERes. Copyright decisions would be made and that file could be utilized until its copyright status was resolved.

Another decision was to only add hard copy items to a course reserve when we had the physical item in our hand. In other words, if a faculty member requested that a book, DVD or video be placed on course reserve, we located the item in our collection, brought it to our processing area and then followed the steps to add it to the course record. If the item was not located or checked out, we followed our searching or recall procedures and when the item was found or returned, we added it to the course. If a request was for an item that we did not own, we submitted a rush order to our acquisitions staff with a note that it was needed for course reserve. When the item arrived, it was rushed through cataloging and delivered to the reserve staff for processing.

One of our most challenging procedural decisions concerned linking to physical items in our collection, i.e., creating hard copy reserve records. The library's OPAC search result display includes the item's location, call number and status which is either "available" (not checked out) or shows a date which means it is checked out and due on the date shown. We wanted items on course reserve to display this same information and that meant editing item records so their location and item type matched their status in the course reserve collection.

We reviewed existing procedures for creating and editing bib and item records in our previous course reserve system. Many of the steps already documented for that system were ones that needed to be incorporated into new procedures. We developed a new template that facilitated adding bib and item records for course reserve and wrote procedures for adding these to the library system. Additionally, we wrote procedures for placing those items on hard copy reserve, i.e., items that circulate. Reserve staff members were trained in these procedures and once we experienced several sessions of adding and editing records, the procedures were finalized and put into place.


## *WORKFLOW*

The reserve staff was accustomed to and trained to working primarily with either electronic or traditional reserve items in our previous system. Two staff members had been the primary contacts for electronic reserve and one worked with traditional reserve although the electronic reserve staff was also familiar with procedures for placing hard copy or physical items on reserve. Implementing ERes made it possible to train reserve staff to handle all formats and to be comfortable working in that environment. This change in workflow made us more efficient and effective as an individual staff member could now process an entire course reserve list.

Each staff member developed his or her own individual workflow patterns but in general followed the same steps. The faculty member initiates the process by bringing a course reserve request and that prompts staff to create an account in ERes and to create the course web site. Hard copy or traditional items would be searched in the OPAC, items retrieved from the stacks and bib and item records created or edited when necessary. These items could then be added and linked to the course site. Copies of articles or book chapters would be scanned or faxed and then linked through the ERes document linking process. As course sites were set up and documents added, reserve staff checked the links to ensure they were functional. If there were any external links that needed to be created such as to the faculty member's web page or other web page, those would be established.

Documents requiring copyright management can be identified as such in ERes through use of the Copyright Management screens. Basic copyright information is added to the document and its disposition is noted. We make use of the Docutek link to the Copyright Clearance Center to request permission for use. If an immediate approval is not available, the document's status remains pending. A document may be made available pending its final copyright status and we chose to make those available. The document's final status is determined by checking the Copyright Clearance Center and then the document entry can be updated. Depending on the decision, the document may remain displayed or removed from the course.

All copyright functions are easily accessed in ERes and a history is maintained for each document added. The Copyright Functions menu option lists all documents in your system in alphabetical order and provides the history, status and fees paid, if any. Copyright reports are easily generated and provide access to usage fees per document, paid items, total amount charged and paid, to name a few. Access statistics are also available as is information about the publishers *or* rights holders associated with your documents. We are using some of these reports already and have plans to explore the others.

ERes provides some basic, easily compiled statistics. We have been mostly concerned with getting the service up and running and only have a couple of month's worth of statistics. In addition, these early access counts contain many that resulted from our testing of the system. Nevertheless, compiling statistics is quick and easy. By selecting the statistics module, we see the "Quick Summary" report that details faculty/non-faculty accounts, active/archived/external courses, copyright/non-copyright documents and free storage space. It is interesting to note that after implementing ERes for the fall semester we have 285 active courses and 207 faculty accounts. Those numbers are much higher than our last count from the previous system. The other reports can be limited to specific time frames and include counts of home page views by month, course page views by month and course page views by month per course. These counts are more pertinent and much easier to compile using ERes.

## TECHNOLOGY CHALLENGES

The technology utilized by ERes can be daunting when explored to the fullest, most especially for libraries used to a more manual reserve system. To fully utilize ERes it is crucial to have access to a reliable scanner and scanning software, computer resources capable of handling large files, fax capabilities and a staff trained to use them. We had already worked in an electronic environment with course reserve and therefore were adept at using the technology available for use with ERes.

Our previous routines set up for scanning for our former reserve process were modified and tailored specifically to ERes requirements. Scanning software was already familiar to the staff and required little modification for its new use. The staff already had a high level of familiarity with computer routines (actions) such as saving and browsing for files, locating and adding urls to records and editing bib and item records. Given that the staff was already skilled, our transition to ERes and mastery of the technology required a minimal amount of time. For libraries starting out with course reserve, ERes will require training and building of computer skills but that time requirement will be less than with other course reserve systems.

## CHALLENGES FOR THE FUTURE

Our major challenge will be to develop the role that faculty can or will play in creating or maintaining their course web sites on ERes. We currently have structured ERes so that all work is done by staff members and faculty members have no access to their course web sites other than viewing. We plan to continue in this fashion for at least the next six months so we can make a more informed decision about the role we want faculty to play. Thus far we have had no requests from faculty to do anything more than what they can do!

Another intermediate step in faculty involvement would be to allow faculty to add material to their course site but in such a way that it is reviewed first. ERes can be set up to allow faculty to add content to their sites but to make it unavailable for viewing until it is approved by a staff member. This allows faculty to add content but still ensures that documents are of good quality, links are properly built and that copyright is properly managed.

In conclusion, we realize that providing reserve services is still an ongoing process. Nevertheless, we feel we are closer to our ultimate goal of providing easy access and use of reserve material by our patrons. Moreover, with ERes the "magic" of making this happen has become easier for staff. ERes has reduced staff processing steps and time required to bring a requested reserve online. We are more satisfied, too, with the integration of our traditional and electronic reserve items into a single interface.

## REFERENCE

Cody, Sue Ann, Pfohl, Dan, and Bittner, Sharon (2001). Establishing and Refining Electronic Course Reserves: A Case Study of a Continuous Process. *Journal of Interlibrary Loan, Document Delivery & Information Supply*, 11(3), 11-37.

# Penfield Library
# Electronic Reserves Initiative:
# A Primer for Electronic Reserves Service

Andrew Urbanek

**SUMMARY.** The State University of New York College at Oswego's Electronic reserve program has been established as a successful highly popular service offered to the university community. The following article chronicles the birth pangs of the service and provides an example of a Docutek supported Electronic reserve service implementation and first year operation. *[Article copies available for a fee from The Haworth Document Delivery Service: 1-800-HAWORTH. E-mail address: <docdelivery@haworthpress.com> Website: <http://www.HaworthPress.com> © 2004 by The Haworth Press, Inc. All rights reserved.]*

**KEYWORDS.** Docutek, ERes, Electronic reserves implementation, Northern New York Library Network, NNYLN, College at Oswego

## *INTRODUCTION*

Many students today enter institutions of higher education as computer and Internet veterans. In response, it is necessary for libraries and

Andrew Urbanek is Senior Assistant Librarian, Penfield Library, State University of New York at Oswego, Oswego, NY 13126 (E-mail: urbanek@oswego.edu).

Docutek and ERes are registered trademarks of Docutek, Inc.

[Haworth co-indexing entry note]: "Penfield Library Electronic Reserves Initiative: A Primer for Electronic Reserves Service." Urbanek, Andrew. Co-published simultaneously in *Journal of Interlibrary Loan, Document Delivery & Electronic Reserve* (The Haworth Information Press, an imprint of The Haworth Press, Inc.) Vol. 15, No. 1, 2004, pp. 43-64; and: *A Guide to Docutek, Inc.'s ERes Software: A Way to Manage Electronic Reserves* (ed: James M. McCloskey) The Haworth Information Press, an imprint of The Haworth Press, Inc., 2004, pp. 43-64. Single or multiple copies of this article are available for a fee from The Haworth Document Delivery Service [1-800-HAWORTH, 9:00 a.m. - 5:00 p.m. (EST). E-mail address: docdelivery@haworthpress.com].

other information centers to adapt to the evolving demands of these students. The level of Internet skills and the heightened expectations of newly enrolled students are forcing information professionals to retool their services to meet these new demands. Full-text databases, virtual reference services and electronic reserves are examples of these 21st century revisions of traditional service.

The State University of New York at Oswego's Penfield Library implemented an Electronic reserves service during the summer of 2002. The following article is intended to describe our implementation process and to provide a primer for other libraries as they move to an electronic reserve service.

## BACKGROUND INFORMATION

The State University of New York at Oswego had a total Fall 2001 enrollment of 8,407 students, 1,203 FTE faculty and staff and 533 part time faculty and staff. Penfield Library had over 100 courses with items on paper reserves. Reserve transactions for the year were 18,661 total circulations. In contrast, in 2002 the Reserves service provided 23,269 circulations for the same number of courses.

The question of what Electronic reserve system to implement was answered when the Northern New York Library Network (NNYLN) generously offered to fund and host the Docutek ERes module for Penfield Library. Northern New York Library Network currently hosts six-area college Electronic reserves services (including SUNY Oswego) on a central server located at their headquarters in Canton, NY.

A single clerk has traditionally processed Penfield's reserves. With the arrival of Electronic reserves as a new service, we estimated that demand (and therefore workload) would increase dramatically. The Access Services Librarian therefore shared responsibility for the processing of reserves as well and initiated a division of labor in the department. The Reserves clerk managed the direct processing and day-to-day maintenance of Electronic reserves while the Access Services Librarian provides oversight for copyright issues and faculty liaison responsibilities.

## OBJECTIVES

As a first step in implementing Electronic reserves at Penfield Library, the following goals were established for the project.

## Limited Pilot Program

It was decided that working with selected technically perceptive faculty during the summer 2001 session would allow greater flexibility in establishing the service. Three faculty members would be able to articulate problems clearly, as well as relate well to any technical "birth pangs." In addition to providing electronic reserves for these professors, paper duplicates were also put on traditional reserves to account for the possibility of a massive problem. Luckily, these "disaster copies" were never needed. The Docu-Fax faculty fax submission module was disabled, since we decided that full control over these documents was crucial. The department wanted to see the entire process firsthand, both in terms of learning the workflow as well as tracking copyright.

## New Formats for Reserves

Rather than simply increase accessibility for traditional print reserves, we wanted to take advantage of the Docutek system to encourage faculty to rethink what kind of items they put on reserve. Through Electronic reserves, images, power point presentations and other new options were possible. Increasing the versatility of the service would in turn increase use of the service. Increasing use served to benefit faculty, students and the library as a whole.

## Remain Within the Safe Harbor of CONFU

As we began exploring this new service, we also did not want to inadvertently make any mistakes that would put our library, service and reputation in jeopardy. As such, Penfield Library follows the CONFU guidelines on what constitutes Fair Use. Although these accords remain unsigned, we followed the University System of Texas, Cornell and other high-use Electronic reserve departments. CONFU places the following limitations on Electronic reserve materials:

- No more than 10% or one chapter from a book
- No more than one article from a periodical issue or newspaper
- No more than one short story, short essay or short poem regardless of collection or anthology
- No more than one chart, diagram, graph, drawing or picture from a periodical, book or newspaper
- No more than 10% or 3 minutes from a motion picture
- No more than 10% or 30 seconds from a music recording or video.

### Bandwidth Sensitivity

A priority for our service has been to accommodate those students who may be relying on dial-up services for connectivity. Luckily, Docutek ERes software calculates download times for both high bandwidth and dial-up patrons. Using this information, we set an in-house guideline of no more than 2 minutes download time for dial-up patrons. Documents that would exceed this limit were divided up into smaller files.

### Platform Sensitivity

Equal access idealism demanded that we mandate that all submitted both PC and MacIntosh computers could view electronic reserve materials alike. No Microsoft dependant files would be accepted. Instead, they would be converted to PDF or other platform-independent file types, where possible.

### House Cleaning

After surveying print copies of high-use reserve items that would be placed on Electronic reserve, we quickly determined that fresh copies were often needed. Some faculty had reused the same paper reserves for years, regardless of wear and tear. As items were going to be scanned into PDF format, we decided to take advantage of this new initiative to request clean copies of course reserves from our professors. These new copies would provide a better scan. They would also provide new clean paper copies for traditional reserves. In a few cases, the need for new reserves challenged faculty to revise their reserves items in terms of currency and frequency of use.

### Tracking

Of all the concerns, tracking our service was paramount. We asked faculty participating in the pilot project to pass paper surveys out to their students, which would be returned to us at the end of the summer session. These surveys would be used to address problem areas before opening the service to the campus in the fall. Our Webmaster assisted us by creating an on-line Electronic reserve feedback site (http://www.oswego.edu/library/circ/feedback.html) that would help us collect responses. A copy of these surveys can be found in Appendix 1.

## Advocacy

Following the 'beta test,' we wanted to make sure that all university faculty would have the option of using Electronic reserves. Librarians were asked to report to their constituent departments and the reserves staff created a brief information packet to help in this regard. An article appeared in the library newsletter. We discovered that new faculty members were especially receptive to embracing Electronic reserves and library orientation programs for these new faculty highlighted an Electronic reserves session.

## Archiving

Initially, we were nervous about using a server that was not directly under our own control. To keep from losing work, we decided to maintain our own archive of Electronic reserve materials on CD-RWs. These archives would remain in the reserves offices and would not be available for circulation. Instead, should the server in Canton fail, we would have backup files to remount on another server. Thankfully, the server has yet to suffer any major problems, so this archiving may be a redundancy but safe is better than sorry.

## Database Linking

Our director provided a limited budget for paying copyright clearance costs. In an effort to use this budget sparingly, we decided to investigate linking to our subscription full-text databases. If a database provides permissions for Electronic reserve PURL-linking, we take advantage of it. Currently, we allow linking to 4 full text databases, but due to licensing restrictions, we remain unable to link to such databases as Lexis-Nexis.

## PILOT PROGRAM

The summer semester at SUNY Oswego is divided into four short semesters. When we were selecting possible candidates for the Electronic reserve pilot project, we decided to try to arrange as little semester overlap as possible in classes utilizing Electronic reserves so we could dedicate our time to testing one class at a time.

As noted earlier, we chose three faculty based on their technological expertise and library familiarity. The first professor taught an education class in conjunction with one of our librarians in the building, the second is the director of the distance-learning program on campus and the third professor had previously established a good working relationship with the reserves department as well as being a computer science instructor. All agreed to help with the pilot project.

### Findings

The summer pilot project was a "learn by doing" experience for our department. In coordinating the division of labor, Electronic reserve processing and copyright clearance, we began to develop a workflow that would eventually be refined into formalized policies and procedures. However, before we could get to that stage, we needed to smooth some of the bumps we encountered during the summer pilot project.

With the help of the three beta-test classes, we solved the following problems:

- Confusion on reserve indexing: Initially, we used our online catalog (Aleph) to index traditional course reserves while we used the Docutek ERes system to index the Electronic reserves. Students reported that they were unsure where to look for the different formats. We erred by assuming students would recognize the different types of reserves and would follow the appropriate links from our library homepage accordingly. Our solution was to index all reserves in the Docutek system. This meant changing the links in our on-line Ex Libris' Aleph catalog to point to the Docutek system. This was easily accomplished with assistance from our Systems Librarian.
- Problems with linking to databases: Almost immediately, we received complaints that students were not able to access full-text documents from our licensed databases, after we had put a link in the Electronic reserve course page. Curiously, only some students were having this problem, while others were able to connect easily. When we realized that only off-campus students were having problems, the diagnosis was easy. We had forgotten to include the campus EZ-Proxy address in the links. Off campus users were not authenticating correctly. Once we fixed these links, another problem was solved.

- Campus traffic: Two Internet pipelines serve SUNY Oswego. The main Verizon line provides the majority of our Internet bandwidth. In addition, an AT&T line provides secondary connectivity. Students, who were on campus and accessing Electronic reserves, were encountering a long delay in download times. Library staff experimented and experienced similar frustrations. We contacted the systems administrator at Northern New York Library Network, who helped us discover the problem by tracing the hops the data took between his server and our students. The culprit turned out to be the AT&T line. For whatever reason, all traffic from Northern New York Library Network to us was being routed through the secondary AT&T line. By working with our Acting Systems Librarian and Campus Information Technologies department, we managed to get the Electronic reserve traffic rerouted through the main Verizon line.

- Lack of feedback: Submitted student paper surveys were not as numerous as we had hoped. In an effort to garner more feedback, we initiated the online feedback form. While we did get some increased response, it was not significant. Feedback continues to be an issue that we are still addressing today.

- Increased usage: We quickly discovered that by using the Docutek system, usage statistics increased dramatically. During our first month of service following the pilot project, we recorded 1586 hits.

- Service confusion: Our pilot project faculty had some difficulty understanding how the service worked, what types of items could be put on reserve and what were the guidelines for copyright usage. In response, we created a Faculty Reserves Contract, which can be found in Appendix 2.

### Workflow

The pilot project identified issues and problems that needed to be addressed before full implementation, but it also gave the department an opportunity to develop a natural workflow. Procedures were created at point of need. These procedures were refined throughout the course of the pilot project and continue to be refined today. It is critical to acknowledge that the project is never done and that changes will and must be made. In some ways, even as we are entering our second year of the service, Electronic reserves remain a pilot project.

In talking with other university librarians who are establishing an Electronic reserve service, we have found that confusion surrounds a systematic process. How does it work? What follows what? Who does what? The following is Penfield Library's Electronic reserve procedure, from reserves submission to end of the semester clean up. While this is a working document, it represents the latest incarnation of Penfield's workflow.

## ELECTRONIC RESERVES PROCESS

### Submission of Faculty Request

- Faculty must fill out a copy of the Course Reserves packet, available on the library website at http://www.oswego.edu/library/circ/ reserves_packetPDF. Included in this packet are reserve submission forms, the faculty contract, a "how to access" instruction sheet and an original copyright materials form. These packets are also available at the circulation and reserves desks.
- Faculty members are responsible for providing copies of materials to be placed on Electronic reserve. Reserves staff are not responsible for this duty. Copies should be clear and legible. The quality of copies deposited is directly proportional to the quality of the Electronic reserve file. Original materials are not accepted; we require photocopies. These materials must be accompanied by full bibliographic data, in the event that items need to be cleared by the Copyright Clearance Center. This information should be filled out on the submission form. If it is not, the form is not complete and could delay Electronic reserve processing.
- Faculty should complete the Electronic reserve form as well as signing the Faculty Contract agreement section and submit it with their materials to place on Electronic reserve.
- Staff members who take the requests and materials must be sure to check that all the information has been filled out and that all the materials submitted are accounted for. They initial the reserves form appropriately. This is crucial in processing Electronic reserves properly.
- Electronic reserves staff checks to see if submitted articles/items are available in online full text databases that allow direct linking. The Reserves Manager is responsible for providing this list to re-

serves staff. When possible, link to the full-text article in the database, rather than create a PDF.

- A maximum of one week is required for the reserves staff to process the Electronic reserves. This does not include the time needed for copyright permissions to be obtained (if necessary). However, items can be uploaded while permissions are being sought.
- Penfield Library reserves the right to deny materials that either violates copyright or prove to be too costly to obtain copyright clearance. At this point, the library is paying for copyright clearance for our faculty. However, should Electronic reserve use increase (forcing copyright clearance requests to increase as well) this arrangement may be reworked.

## Processing Electronic Reserves

- Reserves staff address each request for Electronic reserves in the order in which it is submitted. Requests for immediate processing are considered on a case-by-case basis and only honored under extreme circumstances.
- Electronic reserves staff must determine if a submitted item can be found on a full-text database, before scanning the item into a PDF. If the database allows for linking, they will create a link rather than an entirely new document. When creating this link, the EZ-Proxy protocols should be included in the URL.
- Reserves staff scan items into a PDF format, paying special attention to maximize clarity, reduce unneeded border flash and minimize file size.
- These PDFs are stored on a Zip disk, for later retrieval by the Reserves Manager. The Reserves Manager is responsible for creation of the course account and setting access passwords for Electronic reserve materials. These passwords are securely stored in a private area. Under no circumstances should Electronic reserve materials be created without password protection.
- Reserves Manager is responsible for ascertaining if a submitted article falls within "Fair Use" standards. If the item is Fair Use, it must be clearly marked as such within the ERes system. If the item falls outside Fair Use stipulations, it must be cleared for copyright through Copyright Clearance Center (CCC). If the fee is over $35, the Access Services Librarian must approve the request. The Access Services Librarian will consult with the faculty member to de-

termine if the fee and item is appropriate for the Electronic reserve collection and if not, will provide alternatives.

- Note: The library will *not* fund items from semester to semester that are not being used.
- If an item is not immediately cleared via CCC, notice is sent to the professor that copyright permission has been sought. If permission is denied, the item will be removed. If the item is removed, a notice is placed in the course listings informing the students of this change. Additionally, a notice of denial is sent to the professor.
- The Reserves Manager is responsible for reviewing the PDF files, in the event that a student assistant was tasked with the scanning of the article. If needed, the manager will edit the file for clarity. When reserves staff has created the PDF file, the reserves manager will upload it to the Electronic reserves site.
- The Access Services Librarian will input all copyright data for the item. This process includes seeking (and paying) permissions through the Copyright Clearance Center.
- After the file has been uploaded, reserves staff are responsible for testing the file. Only after the file has been successfully downloaded will the Reserves Manager return the original documents as well as passwords to the submitting faculty member. A "how to access" form is filled out for the faculty member to photocopy and distribute to their students. In no other circumstance may staff divulge course passwords to students.
- The file is then burned to a CD-RW for local storage purposes. These CDs are not available for anyone outside the reserves functions purview. They are stored alphabetically by professors last name.
- Faculty are made aware that reserves staff require a minimum of one week (5 business days) to post Electronic reserves items on a secure server.
- Reserves staff must be alert in reporting problems and/or student complaints to the Reserves Manager.

### Accessing Electronic Reserves

- Only students registered for a particular course may have access to that course's readings, as per CONFU.
- Reserves staff may not reveal course passwords. This is in the purview of the faculty member running the course. If a student re-

quests assistance with their Electronic reserves passwords, they are directed to their professor.

### End of Semester Clean Up

- At the end of each semester, all files are either removed from the Electronic reserves server or moved into an inactive (and protected) directory. The Reserves Manager is responsible for this.
- Reserves Manager may keep the class web site available, including all non-copyrighted material, on a case-by-case basis. After materials have been removed from the course web site, reserves staff will send appropriate notification to the faculty member.
- Clean up begins on the final day of classes for the semester. Semester statistics are computed during the same interval. Any requests to repost the same Electronic reserves for the following semester *must* have Copyright permissions re-established for the following semester, as we interpret the spontaneous use to equate with a single use.

### FACULTY CONTRACT

The purpose of this document is to assist faculty members using Electronic reserves to understand the service. It details not only the rules and processes for accessing their reserves, but also copyright guidelines. This document is one that has undergone changes, prompted both from within and outside of the campus environment. A copy of this form follows in Appendix 2.

The form was originally intended to be an informational document. However, after some miscommunication between faculty and library staff, we included the faculty signature on the submission form to ensure that professors understood the parameters within which Electronic reserves function. By signing this form, they are indicating that they have read, understood and will abide by our rules.

One issue that arose was that some faculty members were reluctant to spend time inputting the bibliographic data into the appropriate field. They reported that they did not have the time. As a shortcut, we allowed these faculty members to submit a copy of the verso page from the journal or monograph. If they were submitting copies of an article, they could add a copy of the verso page to the back of the article, which we would use for copyright purposes. This has worked as a compromise between the library's needs and faculty needs and is an example of the flexibility and responsive attitude needed to make the service flourish.

The faculty contract is something that is continually updated, as both our service, faculty requests and copyright law evolves. Additions and modifications are always being considered. An example of an addition that was made to the Faculty Contract packet was the inclusion of an "original materials" form. This is intended to protect student work that faculty include as part of their course readings. This form asks the faculty member to secure written permission from a student author to use his or her work. Another example was the addition of a line that states that all consumable materials will automatically be denied. In this case, "consumable materials" means commercial worksheets, practice tests or any other material that is designed for one use and then to be discarded. This also includes materials that the faculty member themselves did not create. Both of these changes were prompted by discussions of copyright carried on over various e-mail listservs.

Changes to the faculty contract represent changes to the service. We build our service upon the basis. The Docutek module has a function where all faculty, staff and administrators using the system can receive mass messages to their designated e-mail accounts. We often use this service in conjunction with library newsletter to keep faculty up to date.

A complete Faculty Reserves packet can be found in Appendix 2.

## COPYRIGHT POLICY

When Penfield Library began its service, we were more interested in the creation of a workflow, relevant documents, procedures and timetables than in dealing with the quagmire of interpretation that is inherent in copyright law. As such, we decided to begin our service adhering to the CONFU stipulations. In this way, we decided to remain within the safe harbor that CONFU provides while we concentrated on developing day-to-day aspects of the service.

We also included warnings to faculty that while we were currently supporting copyright permission fees, we reserved the right to deny Electronic reserve items that were too costly to justify. In these cases, we always allow the faculty member the option of contacting his or her department library liaison and requesting that a copy of the item in question be purchased (usually a rush order) or that they place a personal copy of the item on traditional reserves for their students.

The practical workflow regarding copyright processing in our department is easy to manage. When the Reserves Clerk has an article that does not satisfy CONFU or Fair Use, he places the item in a box labeled "Copyright Questionable." During the day, the Access Services Librarian checks the box and decides which items need to be processed for copyright. As the Access Services Librarian uses the Copyright Clearance Center web page to secure copyright permissions, the Reserves Clerk continues to scan and upload the article to the course page. We do not penalize students while we are securing copyright permissions. However, once we are either denied copyright or refuse to pay the royalties, the item is removed from the course site immediately. The Docutek module allows us to create text files on the fly, which we use to put a notice directly in the course page that the document was removed due to copyright issues and that students should contact their professor for more information. A notice is also sent to the professor, detailing options that he or she can take to remedy the problem. Appendix 3 contains such a response.

## *STUDENT FEEDBACK*

Student feedback continues to be one of the challenges facing the department. In the fall 2002-Spring 2003 time span, only five feedback forms were submitted electronically and no paper forms were submitted at all. These forms were all accolades. While this is encouraging, it does not give us any direction to go in improving the service, as well as woefully under-representing our constituent users. It may well be that while we as a staff are excited to offer this new service, for the students it may just be a normal evolutionary process. Regardless, garnering more response from our student users remains a priority in the department.

## *OTHER CHANGES/NEW DEVELOPMENTS*

To give readers a sense of how rapidly things change in the Electronic reserves department, in the course of writing this article (approximately 2-3 months), we have encountered and instituted new changes to the service, detailed including:

1. *Wrong Expectations.* In our traditional paper reserves, we had long ago noted that faculty members continue to re-use the same documents from semester to semester, from year to year. We had anticipated that this practice would continue with Electronic reserves. The first two semesters of the service demanded an intense amount of scanning and processing as faculty signed on for the service. We kept telling ourselves that this would be the only two semesters with such a workload. We assumed that afterwards there would not be nearly as much scanning, since we would naturally be reusing many of the same documents. However, this was not the case. While there was some amount of reused documents, we also noticed new documents, new courses and new faculty using the service. These "newbies" continue to necessitate a high demand for new scanning and processing. This caught us somewhat off guard.

2. *Floating Student Worker.* In an effort to help meet this increased workload, we hired a new student worker specifically for Electronic reserves. Since we did not need all of the student's time, we split his time with the Interlibrary Loan and Government Documents departments. The student was put in charge of mailings, initial document enlarging, photocopying, etc. We have found that having the student worker available to do some of the easiest tasks has given professional staff more time to dedicate to scanning and copyright processing.

3. *One Reserve Form to Rule Them All.* Rather than having to juggle two distinct different types of forms dependant on whether the faculty member was using traditional reserves or Electronic reserves, we merged the two forms into one. This way, faculty need only check off whether they want Electronic reserves or traditional reserves. Penfield Library also no longer offers both print and electronic reserves for a course. Now, it is a case of either/or, not both traditional and Electronic reserves.

4. *Tracking Electronic Reserve Issues.* Keeping track of new software, copyright issues and problems is crucial. In addition to attending any relevant workshops or conferences, the following listservs are highly recommended:

   - CNI-copyright@cni.org
   - ARL-Electronic reserve@arl.org
   - electronic reserves@docutek.com
   - circplus@listserv.boisestate.edu

## FUTURE CONSIDERATIONS

We are never finish updating, tinkering or experimenting with new measures to keep Electronic reserves viable for students and faculty. In addition to responding to student and faculty feedback, the Electronic reserves department closely tracks listserv dialogue and related conferences. In this way, we work to keep proactive pressure on ourselves rather than simply reacting to complaints.

One goal that we have set is to minimize the size of the files we create through scanning. The smaller the file, the less time it takes for students to download and, the less storage space we use on the server. This translates to less storage space faculty and students use saving documents to their own systems. The department is beginning to investigate new data compression methods and software.

Another goal is to implement Optical Character Recognition for our Electronic reserves files, so we can better serve our disabled student community. Using OCR software, for example, students with impaired eyesight could have the reserve items read to them from their own computer or transferred to a Braille keyboard.

We are considering completely doing away with traditional reserves, in favor of 100% Electronic reserves service. There are still several factors that we need to overcome to make this is a viable option. Licensing e-books for Electronic reserves and reworking copyright permission budget to account for the increased demand are just two outstanding issues to tackle. Nonetheless, a complete transfer to Electronic reserves would allow us to use physical space to greater advantage, as well as meeting tech-savvy student expectations.

## CONCLUSION

In the first year of regular Electronic reserve service, Penfield Library tracked over 16,000 hits on Electronic reserve documents, which caused our over all reserves statistics to dramatically increase. It is evident through these numbers, as well as through the rapid adoption by faculty, that this is a promising and high-demand service. We are fortunate to be free from having to maintain a local server in addition to our normal workflow, as well.

These policies and procedures have served us well and continue to be adjusted and amended as need arises. They will be of use to new participants in the Electronic reserves adventure.

## APPENDIX 1. Student Survey

### Penfield Library Electronic Reserves Feedback Sheet

In providing this service to our students, we need to be kept apprised of how the system is working. Please fill out this survey and add any problems, troubles or issues you had while using the Electronic reserves system in the final question. When you are finished, return them to your professor. Thanks!

Please circle the response which best summarizes your experience.

**The instructions were clear and easy to follow**

Strongly Agree      Agree      Neutral      Disagree      Strongly Disagree

**The web site was always up and available**

Strongly Agree      Agree      Neutral      Disagree      Strongly Disagree

**The web site was well organized and easy to navigate**

Strongly Agree      Agree      Neutral      Disagree      Strongly Disagree

**Downloading reserves was easy**

Strongly Agree      Agree      Neutral      Disagree      Strongly Disagree

**Downloading reserves was fast**

Strongly Agree      Agree      Neutral      Disagree      Strongly Disagree

**This service is very helpful**

Strongly Agree      Agree      Neutral      Disagree      Strongly Disagree

**If you have any advice, suggestions or criticisms, please indicate them in the space below:**

## APPENDIX 2. Faculty Reserves Contract

Penfield Library Reserves Faculty Contract

The purpose of this document is to pre-empt misunderstandings between faculty and library staff on the issue of copyright as it pertains to Reserves at Penfield Library.

Electronic reserves are a system where course reserves are scanned into a digital format and uploaded to a database. These files, while secured with a password, are available 24/7 for the students to download and use. Students will need to have the Adobe Acrobat Reader loaded onto their computers in order to view the files. This software is generally preloaded onto most machines. If your students do not have it, it is available at no cost at http://www.adobe.com. In addition to scanned journal articles, you may place other types of items placed on Electronic reserves. Such as:

- Web pages
- Original PowerPoint presentations
- Spreadsheets
- Photographs
- Sound files
- Movie files

In order to access your reserves, direct your students to http://eres.oswego.Northern New York Library Network.net, the Electronic reserves link from the Library Homepage (http://www.oswego.edu/library) or from the Reserves link in the Aleph online catalog. Your reserves will be filed under both your name and your department. If you would like, I am willing to give a demonstration (should take about 5 minutes).

Before these files and materials are placed on reserve, they are evaluated for copyright clearance. If they fall within Fair Use standards, then they are immediately processed and uploaded. Fair Use/ Conference On Fair Use (CONFU) standards are as follows:

Fair Use Stipulations:

- The purpose or character of the use; including if said use is of a commercial nature or nonprofit educational purpose. Any use that detracts from the earnings and/or current market value of a work must be denied.
- The nature of the copyrighted work. In non-technical language, what is being scanned and for what purpose.
- The amount and substantiality of the work in relation to the whole. It is possible to be in violation of Fair Use if more than is needed is used.
- The effect of the use upon the potential market for or value of the copyrighted work. Will reproducing and distributing the item affect any future market value?

CONFU stipulations:

1. No more than 10% or 1 chapter from a book
2. One short story, short essay or short poem (no more than 250 words)
3. One article from a journal or newspaper
4. One chart, diagram, graph, drawing or picture from a book, periodical or newspaper
5. Up to 10% but not more than 30 seconds of a music recording or music video

## APPENDIX 2 (continued)

If you are placing item(s) on Reserve that violate the Fair Use guidelines, the library staff will go through the Copyright Clearance Center to seek permission to use the item(s). This process generally has a fee attached and may take a few weeks for some materials. Penfield Library will pay fees (up to $35 per item), on a trial basis. However, we reserve the right to deny items for Reserve based on copyright clearance and permission charges.

Items that are considered "consumable" (work sheets, tests, etc.) are not eligible for Electronic reserves.

Bring your photocopies to the reserve desk, fill out the reserve form, sign the forms and the library staff will have the materials uploaded within the week (5 business days). **We ask that Electronic reserve readings not be assigned until you have been notified that course reserves are ready**. As always, the sooner the materials are brought in, the sooner they can be processed. Faculty will be given the passwords to their materials (for Electronic reserves) and will be responsible for passing that information along to the students. We recommend copying the instruction sheet and distributing it to your students. Library staff will not give out passwords, as we have no way of knowing who is in your class and who is not.

We ask that you deliver an evaluation to your class at the end of the semester. These surveys should be forwarded to Drew Urbanek at Penfield Library.

If you have any questions or concerns, please contact me either by e-mail or phone. Thanks again!

Drew Urbanek
Senior Assistant Librarian
Penfield Library
312-3567
urbanek@oswego.edu

**Penfield Library Reserves Request Form**

Please complete this form and return it along with all reserve materials to the Circulation Desk at Penfield Library. If you have questions, please call the Reserves Clerk at extension 2560. Please allow 5 working days for all Reserves to be completed.

**Please indicate whether these items are going to be placed on Regular Reserves or Electronic reserves by checking the appropriate box.**

Regular Reserves ☐     Electronic reserves ☐

| Professor's Name | Date submitted |
|---|---|
| Course Title | Course prefix |
| Department | Phone Number/ E-mail: |
| Anticipated enrollment | Semester/ Date of removal |

Please sign here to verify that you have read and will abide by the Faculty Reserves Contract document, where necessary.

_____

date: _____

| **Document Title:** *The name of the item as you wish it to be cataloged.* | **Interval:** *Dates to go on and come off Reserve and/ or **loan** period (2 hour, 4pm overnight, 8pm overnight)* | **Citation Info:** *Author, publisher, title, volume, date, pages, database where article was found, etc. This information is needed to secure copyright permissions for your materials.* | **Verify (Staff):** *Staff must initial* |
|---|---|---|---|
| | | | |
| | | | |
| | | | |
| | | | |
| | | | |
| | | | |
| | | | |
| | | | |
| | | | |
| | | | |
| | | | |
| **Document Title:** *The name of the item as you wish it to be cataloged.* | **Interval:** *Dates to go on and come off Reserve and/ or **loan** period (2 hour, 4pm overnight, 8pm overnight)* | **Citation Info:** *Author, publisher, title, volume, date, pages, database where article was found, etc. This information is needed to secure copyright permissions for your materials.* | **Verify (Staff):** *Staff must initial* |

## APPENDIX 2 (continued)

### Penfield Library Electronic Reserves

*How to access your course Electronic reserves*

**What are Electronic reserves?**

Electronic reserves are articles, PowerPoint files, book chapters, (etc.) for your course that have been scanned and made available to you via the Internet.

**Where are Electronic reserves located?**

Electronic reserves are located on the Internet. Point your browser to the following address to access them: http://eres.oswego.Northern New York Library Network.net or use the "Course Reserves" button from the Library homepage.

**When can I access these materials?**

Electronic reserves are available 24 hours a day, 7 days a week.

**How do I access them?**

1. First, your computer will need a copy of the Adobe Acrobat Reader. Computers in the campus labs (including the library) have this software pre-loaded, so they are okay to use. If you need to get this software for your PC, go to www.adobe.com to get the latest version free. If you need help, call Penfield Library Reference Desk at 312.4267
2. In the address bar of your browser, type: http://eres.oswego.Northern New York Library Network.net and hit Enter or click the Course Reserves link on the library homepage.
3. Search for your course, either by Department (English, for example) or Instructor's name. If your instructor has multiple classes, select the appropriate class.
4. Read the information about copyright and click Accept.
5. Click on the particular item you are interested in. You must now enter the password your professor gave you. Without this password, you will not be able to access the Electronic reserves. If you have lost your password, see your instructor. Library staff cannot give you this password. (*If off campus, you will be prompted a second time. Log in exactly as if you were logging into your Oswego mail.*) You may now read it off the screen, save it or print it. The printed copy will appear clearer than the on-screen copy.
6. If you need help, please call the Penfield Library Circulation Desk at 312.2560.

***Professors: Please photocopy and distribute this sheet to students in your class!***

Professor's Name:

Course:

Password:

## Penfield Library Original Materials Copyright Form

*Attach to front page of original work to be placed on Reserve*

Author's Name (printed):

Title of work:

Number of pages/slides:

Date created:

Course Instructor:

Course Name:

Course Code:

Semester:

Today's Date:

I, the author of the attached work, hereby allow this work to be used for the express purpose of academic instruction in the class listed above.

_____

*Author of work*

I, the course instructor, will not use this material in any other endeavor or class. I will not in any way use this material for my own personal financial benefit. I understand that this material can be placed on reserve for this semester only.

_____

Course Instructor

## APPENDIX 3. Denial of Item Notice

Professor <Name>-

I am sorry to inform you that the Copyright Clearance Center has denied copyright permission for your <item title> reading. As such, we have removed this document from your course listing. However, if you wish to appeal this decision, the center has provided the contact information for contacting the rights holder directly. If you wish to pursue this matter independently and receive permissions, please send us a copy of the letter granting you permission and we will bring that reading back up immediately. In the meantime, if you have an original copy of the item, we can place that on traditional reserves for your class. Alternately, I would suggest contacting your library liaison and requesting that a copy be rush-purchased by the library.

I am sorry for any frustration or complications this may cause. Please let me know if I can be of any assistance.

Andrew Urbanek
Access Service Librarian
SUNY Oswego, Penfield Library

# Embracing Fair Use:
# One University's Epic Journey
# into Copyright Policy

Sandra L. Hudock

Gayle L. Abrahamson

**SUMMARY.** The Library's Circulation Department at Colorado State University-Pueblo began using Docutek's ERes system on a limited basis during fall semester 2002. The nuts and bolts of implementing the ERes system were relatively simple compared to the challenge of revising our reserves policy to accommodate electronic reserves. While the staff realized the necessity of tightening up lax practices that become more glaring in electronic format, convincing the university community proved more challenging. *[Article copies available for a fee from The Haworth Document Delivery Service: 1-800-HAWORTH. E-mail address: <docdelivery@haworthpress.com> Website: <http://www.HaworthPress.com> © 2004 by The Haworth Press, Inc. All rights reserved.]*

Sandra L. Hudock is Assistant Professor of Library Services, Access Services/Interlibrary Loan Librarian, University Library, Colorado State University-Pueblo, 2200 Bonforte Boulevard, Pueblo, CO 81001 (E-mail: sandy.hudock@colostate-pueblo.edu). Gayle L. Abrahamson is Assistant Professor of Library Services, Catalog/Circulation Librarian, University Library, Colorado State University-Pueblo (E-mail: gayle.abrahamson@colostate-pueblo.edu).

Docutek and ERes are registered trademarks of Docutek, Inc.

[Haworth co-indexing entry note]: "Embracing Fair Use: One University's Epic Journey into Copyright Policy." Hudock, Sandra L., and Gayle L. Abrahamson. Co-published simultaneously in *Journal of Interlibrary Loan, Document Delivery & Electronic Reserve* (The Haworth Information Press, an imprint of The Haworth Press, Inc.) Vol. 15, No. 1, 2004, pp. 65-73; and: *A Guide to Docutek, Inc.'s ERes Software: A Way to Manage Electronic Reserves* (ed: James M. McCloskey) The Haworth Information Press, an imprint of The Haworth Press, Inc., 2004, pp. 65-73. Single or multiple copies of this article are available for a fee from The Haworth Document Delivery Service [1-800-HAWORTH, 9:00 a.m. - 5:00 p.m. (EST). E-mail address: docdelivery@haworthpress.com].

http://www.haworthpress.com/web/JILDD
© 2004 by The Haworth Press, Inc. All rights reserved.
Digital Object Identifier: 10.1300/J474v15n01_06

**KEYWORDS.** Docutek, ERes, electronic reserves, copyright, circulation

Colorado State University-Pueblo is a regional comprehensive university serving approximately 6,000 students on and off campus. Formerly the University of Southern Colorado, its name was officially changed in July 2003.

The Colorado State University-Pueblo University Library locates the reserve desk in the circulation area. The reserve materials have traditionally consisted of books (both library books and professors' personal copies), photocopies and audio-visual materials. The long-standing reserve policy, dated 11/95, is a revision of an earlier document that is dated 11/84 (University of Southern Colorado, 1984, rev. 1995). Neither document mentions copyright concerns; they focus more on procedures rather than on policy. Both versions are titled "Reserve Book Policy" and include the statement "The University Library does not photocopy materials to be placed on reserve." Both require faculty to fill out a paper form to accompany any reserve materials; the 11/84 policy even refers to the form as "green." This earlier version also includes the statement "Space is not available for a large number of recommended readings." This line was removed from the 11/95 revision, which is somewhat odd because during the interim, our circulation area was moved to a new location that had even less available space for reserve materials.

However, this was balanced by the discontinuance of the bookstore's policy of sending the last copy of a textbook to the library's reserve desk until it received additional copies for purchase. The photocopied material is contained in file cabinets directly behind the circulation desk, filed under the professor's name. Many of the photocopies get dog-eared over time and missing pages become a problem.

Colorado State University-Pueblo librarians had long been enthralled with the possibility of offering electronic reserves. A subcommittee of the University Teaching and Learning Technology Roundtable (TLTR) was dedicated to exploring this possibility in 1999. However, due to budget and staff constraints, no action could be taken then. It was not until 2002 that the library was able to purchase Docutek's ERes program and staff anticipated its full implementation in spring 2003.

The ERes program was funded in the spring of 2002 by a successful Student Technology Fee grant, in which a committee comprised of students, faculty and Deans read and voted to fund various competing internal grants. Before writing the grant, a literature review showed generally positive feedback on the ERes system and staff invited a representative to make a presentation. An aspect of the program that was a particular draw was the Docufax system, which makes the conversion from paper document to electronic document simple and readily available to anyone with a fax machine.

Since the library was in the process of purchasing a new server for its integrated library system and the existing one was operating at its full capacity, ERes was initially housed on Docutek's server in California. Their technical support staff worked with Colorado State University-Pueblo's IT staff to implement drive mapping so that documents were forwarded to the Docutek server. A fax machine was purchased for the circulation area to facilitate the use of Docufax. The only problems encountered during the off-site hosting were the occasional mysterious disappearances of documents from the fax retrieval queue. This occurred because the retrieval queue was being shared with other libraries being hosted by Docutek.

There was minimal use for the first two semesters (fall 2002 and spring 2003), with only seven and eight classes participating, respectively. During fall 2002, course reserves were accessed a total of 386 times and during spring, 500 times. Since there was no copyright policy in place, materials offered were non-copyrighted, faculty produced items such as study guides and old exams. In August 2002, the new server was installed and ERes was transferred to it. At Colorado State University-Pueblo, the server is housed and maintained in the Information Technology Services department. Since ERes has been on the local server, there have been minimal system problems and Docutek has offered excellent technical support as needed. The only access problem so far encountered has been the printing of black pages due to insufficient memory at some reference area computers. Physical program implementation difficulties were strikingly few compared to those encountered in the development and approval of a copyright policy.

The Electronic Reserves Committee is comprised of the Circulation and Access Services Librarians and Circulation paraprofessionals. The Committee initially began to address the issue of copyright in August of 2002. The first prototype of a copyright policy that spawned a total of fourteen revisions in the course of the next year was written based upon the American Library Association's Model Policy on Library Reserves (ALA, 1982) and the ALA's Proposed Fair Use Guidelines for Electronic Reserve Systems (ALA, 1996). Also consulted and discussed were Laura Gassaway's chapter on copyright in *Managing Electronic Reserves* edited by Jeff Rosedale, and Donna Ferullo's "The Challenge of Electronic Reserves." It must be noted that the ALA Model Policy on Library Reserves was created in 1982 "to assist libraries in establishing reserve copyright policies that function within the framework of the copyright law" (Hatfield, 2001) and addresses solely print reserve collections. Further, the "Proposed Fair Use Guidelines for Electronic Reserves," while providing a professionally sanctioned approach, have never been widely embraced since their development during the Conference on Fair Use in 1996. These guidelines state:

> While acknowledging that some institutions may feel free to adopt and implement them, it was decided at the CONFU session on November 25, 1996, that the proffered guidelines for electronic reserve systems would not be disseminated as a formal work product of CONFU. (CONFU, 1998; Crews, 1999)

Nevertheless, Docutek has integrated key aspects of the "Proposed Guidelines" into its ERes system. For example, Docutek allows the (1) full citation to the materials being provided, (2) the display of copyright notice on a course site's introductory screen as described in Section 108(f)(1) of the Copyright Act and (3) a course level password that is "structured to limit access to students registered in the course for which the items have been placed on reserve and to instructors and staff responsible for the course or the electronic system" (ALA, 1996). Additionally, reserve materials on ERes are only searchable by course number and instructor name, not by the works themselves or their authors' names.

The policy's first draft took a conservative approach to Fair Use by requiring that a royalty be paid on an item's first use, as well as subsequent uses in future semesters. At this time, the idea was to share the cost of royalties by having the library pay Copyright Clearance Center's (CCC) transaction fee up to $6.50 per item and make the royalty fees the responsibility of the faculty member. Library staff would search the CCC database, notify faculty of the amount of the royalty fee and then bill the department via interdepartmental voucher. If an item was not listed in the CCC database, a sample letter requesting permission for use would be provided to the professor. Upon reading the excerpt below from an article that warned its readers not to readily give up fair use claims, since there has not yet been a case regarding electronic reserves, the Committee agreed to extend a "first use, fair use" mandate and require royalties be paid only for the continuing use of an item from semester to semester.

> The safest and most conservative approach is to simply pay for the use of everything in the electronic reserves system. However, this last model is philosophically the most difficult for many librarians to accept, since it simply gives up the right to fair use from the beginning. (Melamut, 2000)

Historically, library policy documents have followed the following route for University approval:

1. The Library Council, composed entirely of library staff
2. The Library Board, composed of two librarians and one faculty member from each college and/or school in the university and one faculty senator
3. The Deans' Council
4. The Provost
5. The President

Given the number of institutional channels the policy needed to go through to gain approval, from the beginning the Committee decided it should consult the University's legal counsel to be sure the draft was in compliance with copyright law and to avoid an unnecessary time lag if it were found not to be in compliance after gaining institutional approval. This decision followed advice given by lawyer and librarian Gretchen McCord Hoffman:

> First, include your entity's legal counsel in the process of creating your copyright policy. Essentially, part of your policy will involve interpreting the law; even if your legal counsel is not a copyright expert, it is paramount that counsel be involved from beginning to end. (Hoffman, 2003)

Another area of electronic reserves, or indeed, traditional reserves, requiring legal advice was that of faculty and student authored materials. While inspecting photocopied materials for potential non-copyrighted items to load electronically, the Committee discovered to its horror that at least two professors had placed students' papers on reserve, complete with names and grades. This was in direct violation of the University's student confidentiality policy. The offending items were immediately removed from reserve and the professors notified. This resulted in the creation of an additional release form that must be signed by the students before their work can be placed on reserve. The Committee assumed that there was such a document as a generic Family Educational Right to Privacy Act (FERPA) release statement that could be attached, only to discover that no such document existed. Therefore, the Committee created one that was in the spirit of and addressed FERPA guidelines. They also added a statement to the procedures manual that requires staff to closely monitor submitted materials to ensure that they are accompanied by release forms.

While reading the professional literature on copyright policy, the Committee also viewed other libraries' electronic reserve policies. Numerous policies and procedures from other universities were reviewed. By the time the draft policy was presented to University counsel in January 2002, it reflected ALA's "Model Guidelines for Electronic Reserves" and the prevalent "first use, fair use" interpretation of copyright law.

University counsel suggested that the Committee look at the policies developed by the state's other universities with electronic reserve systems, including the state's flagship schools, Colorado State University-Ft. Collins and the University of Colorado at Boulder. Additionally, counsel gave recommendations for editing the intellectual property release forms for students and faculty.

Upon viewing both CSU-Ft. Collins' and CU-Boulder's policies, the Committee saw that a broader interpretation of fair use was being applied but opted to continue with a "first use, fair use" approach, requiring royalty payment for subsequent semesters' use of a specific item. Following the next draft revision, the policy was submitted to the University Provost and the University President for approval. It must be noted that the Committee sought to create a policy addressing all reserve materials, regardless of format. This desire to develop an inclusive policy only created confusion and substantial amounts of penned and penciled feedback from the University administration. Efforts were made to clarify the policy by limiting its scope and title to address only electronic materials in the June 2003 draft revision. The Committee will develop a copyright policy for traditional reserves. Since the necessary approval bodies such as Dean's Council did not meet during the summer, no further measures could be taken until fall semester.

The revised electronic reserves policy document had been approved by the prerequisite entities until it reached the Provost and the President. They felt that since it affected faculty, it should be forwarded to the Faculty Senate. The Senate sent it to the Academic Policies and Standards Board (APSB) for initial approval.

In September, the Circulation Librarian contacted CSU-Ft. Collins' reserve staff and found that they had based their policy closely on that of the University of Wisconsin's, which takes a broader interpretation of Fair Use. At the same time, the Access Services Librarian discovered an article outlining CU-Boulder's process of policy development: "Four scenarios concerning fair use and copyright costs: Electronic reserves at the University of Colorado, Boulder." Having their decision-making process succinctly illustrated as they presented differing fair use scenarios to the CU-Boulder Libraries Cabinet resonated with the situation of the Colorado State University-Pueblo Committee. The policy was arrived at based on legal and financial analyses. Specifically, the Second Scenario: Broad Interpretation of Fair Use states:

> Both copyrighted and non-copyrighted materials will be accepted for electronic reserves. Fair Use will be claimed without restriction for all items we own. (Austin, 2003)

With this input from the other universities, along with counsel's advice that there was offered "enough comfort in the other policies" and his recommendation to further edit the policy to reflect their lead, the Committee presented the policy along with these late-breaking caveats to the ABSP in late September 2003. The Board agreed to meet again with Committee representatives to look over the revised document in early October and the motion to bring the policy before the full Senate was made and passed by a voice vote.

Colorado State University-Pueblo librarians spent considerable time in its creation of an electronic reserves copyright policy. One is tempted to offer other institutions at the same juncture wisdom on how to avoid such a considerable time investment. However, each institution must create a policy that fits its demographics and unique situation and this necessitates a process that is outlined by Melamut (2000) as follows:

> When designing an electronic reserves system, each library must identify its own philosophy or approach towards fair use and

whether it will defend a more liberal or conservative approach to the digital environment. (Melamut, 2000)

Perhaps it is poetic justice that the time spent on policy development to fully take advantage of Docutek's ERes program has been inversely proportional to its splendid functionality.

## REFERENCES

American Library Association. ALA Model Policy on Library Reserves. In Gassaway, Laura. (Ed.). (1997). *Growing Pains: Adapting Copyright for Libraries, Education and Society.* Littleton: Fred B. Rothman & Co.

American Library Association. Proposed Fair Use Guidelines for Electronic Reserve Systems Revised: March 5, 1996. In L. Gassaway (Ed.) *Growing Pains: Adapting Copyright for Libraries, Education and Society,* Littleton, CO: Fred B. Rothman & Co.

Association of Research Libraries. (1999). *Electronic Reserves Operations in ARL Libraries: A SPEC Kit.* Washington, D.C.: Association of Research Libraries, Office of Leadership and Management Services.

Austin, B. and Taylor, K. (2003). Four scenarios concerning fair use and copyright costs: electronic reserves at the University of Colorado, Boulder. *Journal of Interlibrary Loan, Document Delivery & Information Supply, 13*(3), 1-13.

Colorado State University. Policy for ER (Electronic Reserve) Access to Published Copyrighted Materials. Retrieved January 22, 2003, from http://lib.colostate.edu/access/Electronic reserve/ERcopyrightpolicy.pdf.

Crews, K.D. (1999). Electronic reserves and fair use: The outer limits of CONFU. *Journal of the American Society for Information Science, 50*(14), 1342-1345.

Ferullo, Donna L. (2002). The Challenge of Electronic reserves. *Library Journal,* 7/15/2002 Retrieved August 1, 2002, from http://libraryjournal.reviewsnews.com/index.asp?layout=article&articleid=CA232350. This speaks to FERPA and student confidentiality issues.

Hatfield, Amy. (2001). Content analysis of restrictive publisher copyright policies for electronic reserves. *Journal of Interlibrary Loan, Document Delivery & Information Supply, 11*(3), 81-101.

Hoffman, Gretchen McCord. (2003). What every librarian should know about copyright: part IV: Writing a copyright policy. *Texas Library Journal, 79*(1), 12-15.

Melamut, Steven J., Thibodeau, Patricia L. and Albright, Eric D. (2000). Fair use or not fair use: that is the electronic reserves question. *Journal of Interlibrary Loan, Document Delivery & Information Supply. 11*(1), 3-28.

Pendleton, Laurence. (personal communication to Sandy Hudock 9/19/2003).

Rosedale, Jeff. (Ed.). (2002). *Managing Electronic Reserves.* Chicago: American Library Association.

University of Colorado-Boulder. Policy for Access to Copyrighted Materials placed on Electronic Reserves. Retrieved January 22, 2003, from http://www-libraries. colorado.edu/ps/crc/ERes/copyright.htm.

University of Southern Colorado. Reserve Book Policy. 11/84; rev. 11/95.

University of Wisconsin-Madison. Library Policy for Electronic Reserve Access to Published Copyrighted Materials October, 1996. Retrieved September 19, 2003, from http://steenbock.library.wisc.edu/reserves/copyrite.htm.

# Electronic Reserves, Library Databases and Courseware: A Complementary Relationship

Steven J. Bell
Michael J. Krasulski

**SUMMARY.** This article discusses the evolution of e-reserves at Philadelphia University, a private institution with 105 full-time faculty and several hundred more adjuncts, from a stand-alone, library-controlled service to an integrated, faculty-driven operation. This change was made possible by the advent of ERes, its unique DocuFax module, and the institution's courseware system. Working together, these systems allow the faculty to largely eliminate the need for manual document scanning, increasing the speed, efficiency and degree of faculty control over the e-reserve process. Uploading content from or linking to content in library databases further eliminates scanning. This process is discussed and illustrated in the article. The complementary nature of ERes and course management software, and why both are needed, is a focal point

Steven J. Bell (E-mail: bells@philau.edu) is Director, and Michael J. Krasulski (E-mail: krasulskim@philau.edu) is Coordinator of Reference Services and Interlibrary Loan Librarian, both at Paul J. Gutman Library, Philadelphia University, Philadelphia, PA 19144.
Docutek and ERes are registered trademarks of Docutek, Inc.

[Haworth co-indexing entry note]: "Electronic Reserves, Library Databases and Courseware: A Complementary Relationship." Bell, Steven J., and Michael J. Krasulski. Co-published simultaneously in *Journal of Interlibrary Loan, Document Delivery & Electronic Reserve* (The Haworth Information Press, an imprint of The Haworth Press, Inc.) Vol. 15, No. 1, 2004, pp. 75-85; and: *A Guide to Docutek, Inc.'s ERes Software: A Way to Manage Electronic Reserves* (ed: James M. McCloskey) The Haworth Information Press, an imprint of The Haworth Press, Inc., 2004, pp. 75-85. Single or multiple copies of this article are available for a fee from The Haworth Document Delivery Service [1-800-HAWORTH, 9:00 a.m. - 5:00 p.m. (EST). E-mail address: docdelivery@haworthpress.com].

http://www.haworthpress.com/web/JILDD
Digital Object Identifier: 10.1300/J474v15n01_07

of this article. The implications of this complementary relationship for copyright, licensing agreements, and information literacy are discussed. *[Article copies available for a fee from The Haworth Document Delivery Service: 1-800-HAWORTH. E-mail address: <docdelivery@haworthpress.com> Website: <http://www.HaworthPress.com> © 2004 by The Haworth Press, Inc. All rights reserved.]*

**KEYWORDS.** ERes, Docutek, electronic reserve, Philadelphia University, courseware

## INTRODUCTION

The traditional view of a library's electronic reserve system is that of a standalone system managed solely or primarily by library staff. The system inputs are hardcopy documents provided by faculty members. Library staff members then scan the documents to convert them to electronic format and add them to the system, whether it is a dedicated product such as Docutek's ERes or a library automation system electronic reserve module (e.g., Innovative or Endeavor both offer electronic reserve modules). Students then use the library catalog or electronic reserve system to locate and acquire reserve readings.

This article presents a different view of Electronic reserves. Given the multiple information databases and management systems available on our campuses, Electronic reserves should function as a complementary system that adds value to and draws on the capabilities of other campus technologies.

The technology systems that mesh particularly well with an electronic reserve are the library's commercial databases, EBSCOhost, etc., and the campus courseware product, Blackboard, WebCT, etc. In this article, the authors will discuss how their Docutek ERes system complements the campus-wide Blackboard Courseinfo courseware system and increases, for faculty members and students, the utility of the library's databases. An unexpected outcome of having the electronic reserve complement these other systems is the creation of greater faculty member's awareness about library resources. Shank emphasized the importance of creating that awareness when he stated, "Academic libraries have been all too absent in the design, development and implementation of courseware. As a result, faculty members do not think of integrating library resources directly into their courseware-enhanced courses" (Shank & Dewald, 2003, p. 38). With planning, training and promotion, academic librarians can gain the support of faculty members for maintaining both Electronic reserves and courseware.

## A CASE STUDY OPENS OUR EYES

On our campus, it took a technology failure to enlighten faculty members to just how well these systems work together. In the midst of the spring semester of 2003, our ERes DocuFax system suddenly failed. DocuFax is a component of ERes that allows faculty members to fax hardcopy documents to the ERes reserve system. DocuFax converts the faxed material into PDF documents that the faculty member can then place on their ERes course site. It took two days to determine that a dysfunctional phone line caused the system failure. In the interim, we received calls from several distraught faculty members who had planned to fax in documents in order to assign class readings for that night or the next day. Searching for a temporary solution, we discovered our faculty members were creating additional, unnecessary work for themselves.

For example, one faculty member wanted her students to read a *New York Times* article. Faculty members often discover a typical spontaneous reading. We learned this faculty members would cut out the article, photocopy it in order to get it onto letter size sheets and then fax the photocopied sheets to ERes using DocuFax. While this is an efficient way to rapidly distribute the article to students, it is a process rife with inefficiency, not to mention that a photocopied article from the *New York Times* does not make for the most legible PDF document.

To quickly make the article available to the students we suggested using one of our commercial library databases to find and acquire the text of the article. Once saved as a text file, it could be uploaded to the ERes page, as one would handle any standard file format. The faculty members could do all of this from home or office with no additional support from the library. The finished text file makes for easy reading and printing. This incident made us realize that many of our faculty members were likely overlooking the ease and simplicity with which they could draw on our existing database content to develop electronic reading lists. This was owed largely, we determined, to the lack of faculty members' awareness about library commercial databases, not just their content but how to retrieve specific articles from them. We decided to do something about this situation.

## *WHY DO WE NEED BOTH?*
## *COURSEWARE AND THE ELECTRONIC RESERVE*

When it comes to educational technologies, institutions of higher education are sensitive to inundating faculty members with a plethora of systems and tools that might ultimately cause more confusion than classroom success. That is why the question about maintaining both an electronic reserve and a courseware system seems to be eternally posed to academic library staffs. In addition, "why have both," has in the past been a frequently posted question on the Docutek ERes electronic discussion list. There is a general assumption that a courseware system eliminates the need for an electronic reserve system, but practice demonstrates that the two systems are not mutually exclusive. They can complement each other.

This would probably not be the case if not for one of the more ingenious features offered by the Docutek ERes system. The key is the linking utility. (See Figure 1.) Every single document added to an ERes course site has a unique URL. The URLs are found within the course page management utility identified as "Link To Documents." (See Figure 2.) Once a faculty member knows this, it is convenient to establish links between a courseware site and the ERes readings. The Blackboard system has a utility for creating links to Web sites. The course site owner simply describes the site and types in the Web address. The process for creating the link to the ERes readings is the same. With both

FIGURE 1

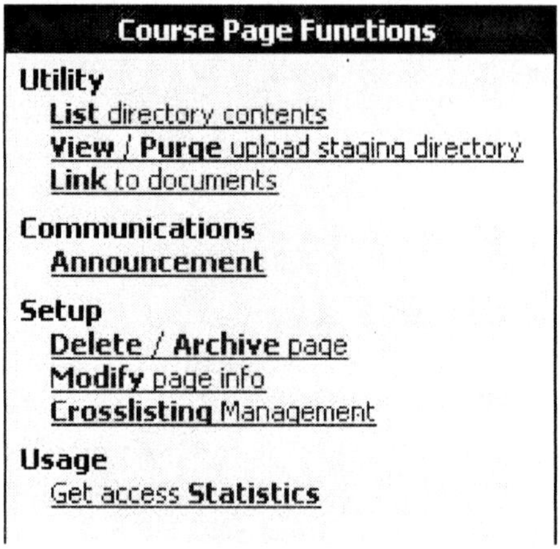

**Course Page Functions**

**Utility**
 List directory contents
 **View / Purge** upload staging directory
 **Link** to documents

**Communications**
 **Announcement**

**Setup**
 **Delete / Archive** page
 **Modify** page info
 **Crosslisting** Management

**Usage**
 Get access **Statistics**

FIGURE 2

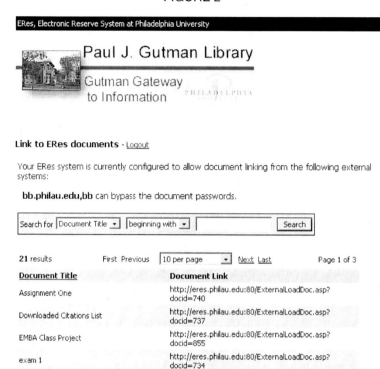

systems running simultaneously, faculty members simply cut the URL from ERes and paste it into the Blackboard template for adding a URL. (See Figure 3.)

The result of this operation is a nearly seamless connection from the faculty member's Blackboard site to the ERes document. When a student clicks on the link to the ERes document, there is no need to authenticate again with the ERes course password. When programmed to allow traffic coming from specified URLs, ERes will not prompt users for the course password. However, it will still require users to read and accept the standard copyright statement. (See Figure 4.) The ERes-based reading can appear right in the Blackboard course site or it can open in a new browser window. The student never needs to connect directly to ERes, but it provides the option of doing so for those students that might find it preferable to visit ERes directly to obtain all of the course electronic reserve material.

FIGURE 3

FIGURE 4

View ERes Document: Are We Ready for the Virtual Library

Please be advised that to use electronic material in ERes, you must agree to the following by clicking the Accept button:

The copyright law of the United States (Title 17, United States Code) governs the making of photocopies or other reproductions of copyrighted materials. Under certain conditions specified in the law, libraries and archives are authorized to furnish a photocopy or other reproduction. One of these specified conditions is that the photocopy or reproduction is not to be "used for any purpose other than private study, scholarship, or research." If a user makes a request for, or later uses, a photocopy or reproduction for purposes in excess of "fair use," that user may be liable for copyright infringement.

[ Accept ] [ Decline ]

While it is true that faculty members could add digitized article readings to the Blackboard course site without an ERes system, we believe having both systems, along with DocuFax, allows faculty members or the library staff to eliminate the scanning process. This is a great timesaver for faculty members and the library staff and often spares students the misery of reading poorly scanned articles. We also believe that because ERes is designed for the reserve function while Blackboard is not, ERes offers a far better environment than Blackboard for managing copyrighted articles that are being digitized and posted in an electronic environment. The seamless connectivity between ERes and a courseware system means that students are really required to use just a single system. We receive virtually no negative feedback about the need to remember extra passwords, the need to learn multiple system interfaces or the usual student and faculty members' complaints when the library subjects them to a multitude of electronic resources.

What is less clear is how well this would all work with other electronic reserve systems, specifically those offered as modules within library automation systems. We suspect that it is also possible to identify unique URLs for each document added to the Electronic reserves and that those URLs, once identified, could be added to the courseware site.

However, since those systems offer nothing like DocuFax, they do not have the simplicity and efficiencies of ERes. Given the way that ERes is designed to clearly identify all of the URLs associated with each document added to the course page, faculty members can truly self-manage their own reserve site and the process of creating links between it and their courseware site. Our experience shows that our faculty members master these skills after a single demonstration and in some cases, simply by following our documentation. Faculty members have little time or patience for technologies that are not easy and convenient. We are confident that ERes is both.

## EXPLOITING EXISTING CONTENT: THE ELECTRONIC RESERVE AND LIBRARY DATABASES

The electronic reserve system at the library, where one author previously held a position, followed a more traditional procedure. All faculty members were required to submit hardcopy documents that would then be scanned into the electronic reserve system. Faculty members were generally satisfied except for their inability to post documents electronically. Faculty members begged library staff members to rapidly put materials on reserve. Since that was not always possible, such as on weekends or evenings, staff would frequently field complaints from disgruntled faculty members. At Philadelphia University, we avoid these unpleasant situations by combining the strengths of ERes and our library databases. The dynamic of the reserve function shifts from the library's control to the faculty members' control. Faculty members place their items on reserve when they want, getting material to students quickly and unimpeded by reserve room turnaround times.

The library's commercial databases present faculty members with a vast inventory of potential content, all of which can be added to an electronic reserve site from their desktops without ever touching a piece of paper or communicating with a library staff member. At our library alone, through our aggregator databases and e-journal collections, we offer faculty members over 8,000 electronic publications in full-text format. To exploit the relationship between the databases and Electronic reserves, faculty members must search for, retrieve and capture their materials using the library databases. This presents several challenges, which, if not overcome, will simply move the traditional reserve paradigm from a paper to an electronic setting:

- Faculty members must be able to identify the appropriate database that contains the needed content;
- Faculty members must be able to effectively search for unique articles across a variety of library search systems;
- Faculty members must have sufficient technology savvy to understand the differences between durable links, PDFs and text files;
- Faculty members must know how to transfer captured content into the electronic reserve site.

The major challenge to the library is to make faculty members aware of their ability to do all of this. Initially, it took a DocuFax failure to make us realize that our faculty members were simply bypassing the scanning routine, which is hardly taking advantage of the technological capacities of databases and Electronic reserves. It dawned on us that we could use our faculty member's motivation to manage their own electronic reserve operation and their desire for fast, rapid content loading, to achieve some of the same information literacy goals we were already communicating to our students. At least two ACRL information literacy competencies, accessing needed information efficiently and understanding the economic, legal, social and ethical uses of information, can be applied to the faculty members' environment. Training faculty members to exploit the relationship between the electronic reserve and library databases is a perfect vehicle for elevating faculty members' information literacy skills. This is a win-win situation for the librarians and faculty members because the more knowledgeable the faculty members are about library databases and using them efficiently, the more likely it is that faculty members will become allies in our initiatives to create an information literate student.

Using our information literacy programming as a model, we have developed a presentation and a Web-based tutorial page for faculty members that are designed to help them overcome the challenges listed above. The sessions and tutorial can cover all the possibilities for exploiting the library-electronic reserve connection from:

- Identifying and using durable links in library databases, using database advanced search features to perform searches that retrieve a single, targeted article
- Capturing text (for databases that do not yet offer durable links)
- Saving text or PDFs to a computer
- Loading the captured content into the ERes course site.

Faculty members are not always comfortable coming to group training sessions, so we also provide individualized sessions. Often, faculty members can learn all of the necessary techniques from our "Capturing Content" tutorial (found on the Web at http://staff.philau.edu/bells/capturecontent.htm). After reviewing the tutorial, faculty members more easily grasp the concept of using database content to populate the course reserve page and may only need some limited, additional phone support to accomplish these tasks.

With respect to the database, electronic reserve and courseware triumvirate, there is one more system worth mentioning. Since faculty members need to quickly match a desired journal with an available full-text database, serials management software is recommended to aid in this task. A service such as SerialsSolution or TDNet will greatly facilitate faculty members in their effort to determine which database has the full-text version of a journal.

In addition to the time needed to train faculty members to maximize what ERes, library databases and courseware offers them, librarians need to educate faculty members about the copyright constraints and licensing agreements that affect materials being placed on ERes. The traditional reserve model encompasses all types of media, from textbooks to photocopied journal articles, all primarily for in-house use. Policies and procedures created to comply with copyright guidelines, notably section 107 of Title 17 of the U.S. Code, in the print realm may be different in an electronic world where licensing agreements must also be considered.

In determining how to interpret the fair use guidelines of section 107 when applied to Electronic reserves, the authors look to experts such as Donna Ferullo, Director of the University Copyright Office and Assistant Professor of Library Science at Purdue University. According to Ferullo, a copyright compliant electronic reserve environment utilizes linked or saved articles from a legally licensed database that exists within a password-authenticated environment when posted online. ERes offers just such an environment for delivering Electronic reserves to the campus community (Ferullo, 2002). This mode of delivery is actually more copyright compliant than the copying and scanning practices of traditional reserves.

Both ERes and courseware systems offer password-protected environments. Passwords limit access to those students enrolled in a particular class. Since creating a password protected environment in ERes is optional faculty members need to understand why it is important to set passwords for their reserve materials. Even though courseware does offer the type of environment recommended by Ferullo and it could serve as an electronic reserve, just having courseware alone would mean doing without DocuFax. We also advocate ERes for the superior copyright management features built into the system.

Whenever possible, we recommend that our faculty members use the durable or persistent linking features of our library databases. Electronic reserves and library databases work together in this complimentary way. It is also a particularly significant feature for maintaining copyright compliance because the library's license to the database content provides legitimate access. There is no reproduction of copyrighted article content, just the creation of a bridge from the electronic reserve or courseware to the content. A durable link is the URL of a specific article or periodical within a library database. Once faculty members understand the durable link concept and know how to place them into ERes or courseware, it allows them to quickly create a link for students directly to a course reading. It also further eliminates the need for time consuming scanning.

While we have yet to encounter a database aggregator who opposes our creation of links between their content and our electronic reserve, it may be worthwhile to examine your library's licensing agreements. Linking from databases where licenses allow it is the best route to follow suggests Donna Ferullo. However, she warns, "most contracts from vendors/publishers do not include this provision, but it is certainly a viable and negotiable issue" (Ferullo, 2002, p. 35). If there is any ambiguity in your library's database licensing agreements about the permissibility of linking, re-negotiate them so your rights are clearly spelled out.

Although ERes creates the copyright compliant environment, the burden of maintaining and enforcing copyright after materials are posted on ERes shifts from the library to faculty members. With traditional reserves, the library is responsible for removing reserve materials at the end of the semester. The responsibility for archiving or deleting the contents of their ERes accounts belongs to the faculty member. That does not mean we abandon our faculty members knowing the requirements of fair use. Rather, we provide faculty members with the guidance they need to properly navigate copyright.

ERes changes the roles of librarians and faculty members. The faculty members become reserve managers and librarians become their consultants. Faculty members are empowered to modify their reserve content anywhere at anytime without library intervention. Today's *New York Times* article is easily captured from a library database as a durable link or text and uploaded to ERes to become tomorrow's reserve reading.

Ultimately, the degree to which faculty members exploit the library's electronic reserve and complimentary systems depends on the quality of the library's training programs and supporting documentation.

## CONCLUSION

While there are obvious administrative efficiencies to be gained by combining the strengths of an electronic reserve system with the library's databases and the campus courseware system, there are perhaps less tangible pedagogical benefits that can result from the complementary relationship between the systems that is described in this article. When faculty members have growing access to an electronic teaching environment, along with access to information via the Internet and other commercial information providers (e.g., XanEdu, publisher-based systems such as Pearson ResearchNavigator), the possibilities for the further marginalization of academic librarians grows greater.

From our perspective, the ERes, electronic reserve system, along with DocuFax, is a sound investment because we are able to leverage it to create greater awareness about and use of our library commercial databases, which are actually a far greater investment of institutional resources. The library, rather than facing marginalization, is able to further integrate itself into the teaching and learning process by establishing a presence within the campus courseware system. By design, courseware systems offer no presence for the academic library and faculty members can unknowingly create an information environment for their students that largely ignores library resources. Showing faculty members the benefits of bringing together the electronic reserve, library databases and courseware helps them and academic librarians to mutually achieve the shared goal of connecting students with high-quality information. Nor does it hurt that our faculty members are gratified that we can support their desire to self-manage the growing array of instructional technologies at their disposal without adding disruptive complexity to an expanding universe of educational possibilities.

## REFERENCES

Ferullo, D. L. (2002). The Challenge of Electronic reserves. *School Library Journal Net Connect*, *33*(5), 33-35.

Shank, J. D., & Dewald, N. H. (2003). Establishing Our Presence in Courseware: Adding Library Services to the Virtual Classroom. *Information Technology and Libraries*, *22*(1), 38-43.

# A Consideration
## of Docutek's Electronic Reserve System in a University's Courseware Environment

Donna H. Ziegenfuss
James M. McCloskey

**SUMMARY.** On many college and university campuses, electronic reserve systems often operate concurrently with courseware products. In this article, consideration is made of the pros and cons of such efforts along with a table that compares features of three systems. The table can be used for advising faculty and comparing systems. *[Article copies available for a fee from The Haworth Document Delivery Service: 1-800-HAWORTH. E-mail address: <docdelivery@haworthpress.com> Website: <http://www.HaworthPress.com> © 2004 by The Haworth Press, Inc. All rights reserved.]*

**KEYWORDS.** Docutek, ERes, electronic reserves, courseware

Donna H. Ziegenfuss is Head of the Faculty Technology Center and James M. McCloskey (E-mail: James.M.McCloskey@widener.edu) is Head of Public Services, both at Wolfgram Memorial Library, Widener University, One University Place, Chester, PA 19013.

Docutek and ERes are registered trademarks of Docutek, Inc.

[Haworth co-indexing entry note]: "A Consideration of Docutek's Electronic Reserve System in a University's Courseware Environment." Ziegenfuss, Donna H., and James M. McCloskey. Co-published simultaneously in *Journal of Interlibrary Loan, Document Delivery & Electronic Reserve* (The Haworth Information Press, an imprint of The Haworth Press, Inc.) Vol. 15, No. 1, 2004, pp. 87-97; and: *A Guide to Docutek, Inc.'s ERes Software: A Way to Manage Electronic Reserves* (ed: James M. McCloskey) The Haworth Information Press, an imprint of The Haworth Press, Inc., 2004, pp. 87-97. Single or multiple copies of this article are available for a fee from The Haworth Document Delivery Service [1-800-HAWORTH, 9:00 a.m. - 5:00 p.m. (EST). E-mail address: docdelivery@haworthpress.com].

Digital Object Identifier: 10.1300/J474v15n01_08

Emulating the learning experience of a classroom has been called the "last frontier" in the effort to automate higher education (Cohen, 2001). Many academic institutions have implemented commercially developed portal and course management systems to provide a one-stop computing experience for students, faculty and administrators. For example, students can register for classes, pay bills, click into their course pages on the web for assignments, join threaded discussions, get grades and more. Academic librarians have actively participated in this movement through such means as virtual reference desk services, electronic document delivery, strategic placement of online public access catalog and database links and electronic reserves.

The professional literature shows that a transformation is occurring in higher education that is challenging the traditional beliefs about how, when and where college students learn. These trends are causing reevaluations of institutional purposes and missions (Astin, 1993; Hornblower, 1997). West (1999) states that this new paradigm of teaching and learning will force a shift in how instruction is delivered to students, as well as how the institution conducts its daily business. This article will consider the environment on one campus in which a library's electronic reserve system, specifically Docutek Inc.'s ERes, operates, often overlapping, complimenting and competing with the existing portal and courseware systems.

Traditionally, instructional models were based on the premise that all students bring to college the same experiences, learning goals and styles and the ability to progress at the same rate when learning (Twigg, 1997). These assumptions are no longer considered valid. Faculty and institutions are being forced to evaluate more flexible pedagogical models to address the issues of this challenging and changing diverse student population.

In addition to changing student demographics, rapidly developing technological advancements have resulted in change on a global scale that higher education is beginning to feel more poignantly. Before the Internet, accessing information was laborious and information changed at a much slower pace. Today, knowledge in many fields changes at an exponential rate. Additionally, this massive amount of content and information is constantly changing and evolving and this adds yet another layer of complexity as students try to keep up with the evolving information and technology. The jobs of today require not only accruing content specific knowledge but also necessitate learning a variety of different skills such as people skills, project management skills, collaborative skills, communication skills and a commitment to lifelong learning (Good, 1988).

Institutions and educators are coming to realize that new emerging technologies can provide them with alternative strategies and tools for responding to diverse student learning styles, providing opportunities for active student involvement and expanding access to higher education. Technology from a broader and more educational perspective has begun to change not only how students learn but also when and where they learn (McGrath, 1998). The Internet, web-based learning, laptops, powerful local networks and other evolving wireless technologies are taking the learning outside of the classroom, or bringing resources into the classroom that were not possible just a short time ago. Technology is also sparking a renewed interest in pedagogical teaching and learning strategies because these emerging technologies have the ability to create powerful contexts for learning that are different from the traditional passive lecture format of the past (Kinnaman, 1993). The social learning environment also takes on a new prominence when knowledge creation is thought to develop through the interaction of learners, the environment and association with other learners. Virtual universities, e-learning and distance education are becoming topics of discussion at traditional universities for both economic and student-related reasons. This appeal of being able to extend the reach of learning beyond the walls of a classroom, or even an institution, has possibilities that may be enticing although studies presented in the literature show no significant difference between a traditional course and computer-based course (Russell, 1997).

This transformation of the classroom also affects the relationship between the faculty and students. Technology tools have been found to increase student motivation, promote cooperation and collaboration and encourage constructivist-teaching environments (Jonassen, 1994). Technology tools can also be used to address diverse learning styles, multiple intelligences and create a more authentic learning environment for students (Gregoire, Bracewell, & Laferriere, 1996; McGrath, 1998). The changing character of today's college students necessitate a change in teaching and learning strategies, as well as a redesign of learning environments. If educators are to prepare students for life in the real world, they must address the issues related to the development of technology skills, problem solving strategies, communication skills and critical thinking abilities in addition to providing traditional academic content material. Institutions are reluctantly and cautiously embracing technology at the cost of not wanting to be left behind. Lynch (1998), Executive

Director of the Coalition for Networked Information, confirms this uncertainty for the future when he states, "My feeling is that this issue is going to become even more complex and visible in the coming years as distance education, instructional technology, multimedia authoring and distribution and very sophisticated network applications become ever larger factors in the activities of universities and colleges. Institutions will continue to grapple with how to best organize this growing span of disparate but interrelated activities and services."

Should this uncertainty about the future compel universities to take no action at all when it comes to technology tools and strategies for the fear of making the wrong choice or going down the wrong path? Absolutely not! Due to the exponential change rate associated with technological advancements we will never find the clarity of calm waters into which to jump. The longer an institution waits, the harder it will be to catch up. At Widener University, although we have not totally committed to one institution-wide standardized online learning (course management) tool, a variety of products are used to meet the needs of particular groups of faculty and students. Rather than take no action and have no online learning tools, through the initiatives of different departments, the university has adopted several tools. Although this is not the most efficient and economic way to implement technology tools, faculty and students do have choices and opportunities to experience the benefits of web-based learning tools until such time that the university reaches consensus on a standard web-based teaching and learning tool. The three tools available for faculty for design and development of online course materials offer varying degrees of richness and different degree of usability ease and allow faculty to begin at a level where they feel comfortable (see Table 1).

Courseware tools share a certain core set of basic features (Shank & Dewald, 2003): powerful resource sharing, communication and assessment tools. One web-based tool offering powerful but easy to use resource sharing, is Docutek, Inc.'s electronic reserve system named ERes. ERes enables faculty-initiated uploading of course content to a password protected course page. Faculty can also have material posted to their course page by library staff that scan and digitize material using Adobe Capture software. In addition, ERes allows for creation of links to web pages and full-text database articles, as well as links to librarian created subject guides. ERes also contains copyright management capabilities that many more robust courseware products lack. While primarily an electronic reserves system, ERes also allows students, instructors and librarians to communicate via e-mail, a discussion forum for dis-

TABLE 1. Course Management Tool Comparisons: Which Is Best for Your Needs?

| Campus-Cruiser | Web-Study | ERes | |
|---|---|---|---|
| X | X | X | **Post course announcements**<br>**CampusCruiser and WebStudy**–announcements "filter up" to the main login or portal screen so that students can see an announcement without actually going into the course.<br>**ERes**–students see course announcement after they enter the ERes course. |
| X | X | X | **Load a Syllabus**<br>**CampusCruiser**–You can upload a word doc syllabus just as you do a file attachment in Campus Cruiser e-mail. There is a designated "syllabus location" within the course.<br>**ERes**–You can upload a syllabus that can be placed with the other course files, but ERes does not have a separate "syllabus location" within the course.<br>**WebStudy**–You can upload a full syllabus into the *Materials* section and create a *Timeline* link. However, the *Timeline* (broken down into sessions or classes) is designed to be the working syllabus tool. |
| X | X | X | **Conduct online discussions on the message board or forum**<br>**CampusCruiser**–provides both a flat message board and a threaded discussion view of the message board. Students can access the message board or forum board directly from the *My Assignment* section and the *Message Board* section if the message board is set up as part of an assignment.<br>**ERes**–provides a flat message board option for discussion of documents and readings posted in ERes.<br>**WebStudy**–has a *Forum* section that uses a threaded discussion format. It also provides private team discussion forums that are separate from full class forums and they can be used for team or group projects discussions. |
| X | X | X | **Conduct a live chat in real time (synchronous discussion)**<br>**CampusCruiser**–provides a live chat area accessible to students and/or students and faculty, and the discussion transcript cannot be archived.<br>**ERes**–has a chat function for real-time discussion of articles posted within ERes and the discussion transcript cannot be archived.<br>**WebStudy**–has a *Live* tab that provides chats for courses, teams, and the entire WebStudy community. Faculty can participate in full class chats, but they cannot participate in team chats that are restricted to team members only. Both team and full class chat transcripts can be archived as a text file. |
| X | X | X | **Post Internet links**<br>**CampusCruiser**–has a *Bookmark* section where you can add links and/or categories to organize web links. Faculty can also add links that are directly connected to assignments (separate from the *Bookmark* section) and accessed through the *My Assignment* feature.<br>**ERes**–web pages and links can be posted just like any other file or article. ERes also provides links to course-specific content material available on the Wolfgram library site, as well as access to a virtual reference librarian service.<br>**WebStudy**–has a separate *Web Link* feature for listing links. In addition to the Web Link feature, web page links can be posted as independent live or static pages if they are posted to the *Webstorium* and then associated with assignments and *Timeline* sessions. |

## TABLE 1 (continued)

| Campus-Cruiser | Web-Study | ERes | |
|---|---|---|---|
| X | X | X | **Provides consistent navigation**<br>**CampusCruiser** and **WebStudy**–both tools have consistent navigation schemes and course sections and students see basically the same course menus or tabs for each course. In both CampusCruiser and WebStudy the tabs can be customized for each particular course.<br>**ERes**–the chat and message board navigation menus are consistent for all courses. The location of the course content section is also consistent, but since individual instructors develop the course from scratch, the content folder structure may vary. Faculty can set up course content folders exactly as they prefer which can be considered to be an advantage. |
| X | 1/2X | X | **Post assignments and retrieve student submitted electronic assignments**<br>**CampusCruiser**–faculty have an *Assignment Tool* where they can post an assignment and automatically create an associated message board, student as well as an entry in the grade book. Students can submit assignments electronically through the *My Assignment* student tool and faculty retrieve student assignments in a single location.<br>**ERes**–student assignments can be posted as files inside the ERes course but there is not a specific assignment tool or location for assignments. Students cannot turn in assignments electronically.<br>**WebStudy**–has a *Work-To-Do* tab where faculty can set up assignments that are automatically posted to the Class Calendar and onto the *Timeline*. This feature also permits the faculty member to attach files to the assignment and automatically creates an entry in the grade book. Students can submit assignments electronically through the *Work-To-Do* feature. |
| X | 1/2X | X | **Has a place for sharing of files by students and faculty**<br>**CampusCruiser**–has a *Shared File* section where students and faculty can both post and view files that can be downloaded by all members of the class. Files can be organized in folders.<br>**ERes**–any file posted by the faculty can be viewed or downloaded by any students, however, students cannot post files to the ERes course. ERes does provide a file sharing option between courses so that content can be posted once and then shared from one course to another.<br>**WebStudy**–has a *Materials* section of the course where faculty can post materials for all students to access. Every student and faculty member has space in the *Webstorium* to post their own files online but all users cannot view them unless placed within the course and/or associated with course assignments or sessions in the *Timeline*. |
| X | | X | **See a student class list**<br>**CampusCruiser** and **WebStudy**–Both tools have a class list feature where faculty can view a student class list and access student e-mail addresses.<br>**ERes**–all students access course materials using the same password and therefore there is not an official "class list" of enrolled students. This can be an advantage in that you are not restricted to only valid Widener users and can permit access to visiting colleagues, etc. |

| Campus-Cruiser | Web-Study | ERes | |
|---|---|---|---|
| X | | X | **E-mail students from within the course**<br>**CampusCruiser**–uses the existing CampusCruiser e-mail system. Faculty and students can both e-mail all members or pick individual class members to e-mail from a course list.<br>**ERes**–does not have an e-mail system associated with it, but students can e-mail the instructor from within the course.<br>**WebStudy**–uses a separate internal e-mail system accessible through WebStudy only and instructors can e-mail all students, as well as pick and choose single students from the class list. Note: Students and faculty can post another e-mail address under personal settings, but this is not used for the main course e-mail correspondence. |
| X | | X | **Post student grades in an online grade book**<br>**CampusCruiser**–has a new grade book feature where assignments can be entered into the grade book and then graded using a variety of customizable grading schemes. Grades can be calculated automatically to a letter grade depending on the assigned grading scheme. Faculty can also comment on student work when grading. Grades can be exported as an Excel file. This grade book feature gives students access to their own grades during the semester, but the instructor must still deliver final grades to the registrar in the traditional way at the end of the semester.<br>**ERes**–There is no grade book associated with ERes.<br>**WebStudy**–has a grade book feature that is tied to the *Work-To-Do* tab. Faculty can pick up electronically submitted assignments and grade them there. Faculty can also add comments for students when grading and the grades can be exported to an Excel file that opens directly in Excel. |
| X | | X | **Post assignments on a course calendar**<br>**CampusCruiser**–has a course calendar but the instructor must place assignments manually on the calendar. Students can copy class calendar entries onto their personal calendars.<br>**ERes**–has no course calendar feature.<br>**WebStudy**–has an integrated course calendar where assignments and forums are automatically posted to the course calendar and students can access information by clicking on calendar assignment icons placed on the calendar. |
| 1/2X | | X | **Has a team project feature**<br>**CampusCruiser**–has semi-team functionality through the assignment feature where different groups of students can be assigned different assignments and resources. However, there is not a "location" specifically for students to share teamwork and conduct team-specific activities.<br>**ERes**–has no teamwork features.<br>**WebStudy**–has a *Teams* tab where students can be divided into teams and where they have a special location to share files, have forum discussions and chats. Students divided into teams cannot see the work and projects of other teams and can participate in only their own team-specific chats and forums. |
| | X | | **Document copyright management**<br>**CampusCruiser and Webstudy**–course content is prepared and posted by the instructor and copyright management is the sole responsibility of the instructor.<br>**ERes**–faculty can have articles scanned, digitized, and posted to their courses by the Wolfgram Memorial Library Reserves Department staff. Assistance with copyright management issues is also offered through the Public Services Department of the library. |

## TABLE 1 (continued)

| Campus-Cruiser | Web-Study | ERes | |
|---|---|---|---|
| | | X | **Administer online quizzes**<br>CampusCruiser–has no online quizzing or testing features.<br>ERes–has no online quizzing or testing features.<br>WebStudy–has online testing features so that students can take self-assessment quizzes online. |
| X | X | X | **Additional pros**<br>CampusCruiser–has a single login for e-mail, classes features, and course registration. There are similar icons, menus, and tools so it can be easier to implement since students are already familiar with the CampusCruiser e-mail environment. Faculty can use this tool to integrate the other web-based tools and environments into a single location.<br>ERes–provides flexibility in enrollment and security, assistance with copyright management issues, and responsive campus support through Wolfgram Memorial Library staff.<br>WebStudy–the most robust of tools, contains components such as online testing features that the other tools do not provide. Online 24/7 help, and phone assistance is offered through WHYY. |
| X | X | X | **Additional cons**<br>CampusCruiser–faculty who are already dissatisfied with CampusCruiser e-mail and communication features do not want to even try, review, or use the newly implemented classes' features.<br>ERes–is not designed as course management software, but rather as an electronic reserves tool so it lacks assignment, grading and other classes' features that CampusCruiser and WebStudy provide.<br>WebStudy–although the most robust tool, has the steepest learning curve due to the unfamiliar course environment and is not connected in any way with the student administrative system. |

cussing online documents and a chat feature for online classroom discussions in real time. This tool is used most often by faculty interested in providing web-based online access to course materials and as a way to enhance the traditional face-to-face course. Of the three options available at Widener University, faculty and students report that ERes is the easiest tool to learn and use.

The second tool in use at Widener is a modified course management tool associated with the campus portal, "CampusCruiser." This tool provides the capabilities available through ERes (minus the copyright management of documents) as well as additional features normally associated with web-based courses such as: automatic student enrollment, online assignment tracking and submission, an online grade book, class e-mail functions and the capability for student file sharing. This online course tool, supported and maintained through Widener University's instructional technology department, is integrated with the campus e-mail system and other web-based student services.

The third option available to Widener faculty is "WebStudy," an online course management tool offered through WHYY television, a local PBS affiliate. The University's school for continuing studies for both hybrid and totally online courses primarily uses this tool. These courses are administered through University College and not connected to the main student administrative database. Some departments have now trained their own administrators to manage their own discipline area courses and programs. This tool contains all the web-based course features of CampusCruiser but has an additional online testing and quizzing feature. Due to a costly student per course fee structure, this software is not readily available to all schools within our institution.

Each of these three tools currently offer faculty a variety of feature choices and faculty are free to choose the options that they prefer and the ones that best suit their needs. Some faculty use a combination of several tools to achieve the level of customization or features they require to meet course goals and objectives. The work of administering the three tools is distributed over three different departments and the responsibility of putting course materials online do not fall on the shoulders of one administrator or department. Faculty can use any one of the tools to create a web-enhanced course and supplement a traditional face-to-face class, but their choices and options narrow as they opt to utilize more sophisticated online learning features.

The flexibility offered through the availability of three completely different web-based learning products does not come without drawbacks. Each is administered through a different department, which can cause confusion for both students and faculty. These products have overlapping features, such as synchronous/asynchronous communication tools and online depository that seem redundant and add to overall costs. WebStudy has a separate internal e-mail system that students and faculty must manage and check in addition to the general campus e-mail system. All three systems have separate logins and passwords so that students could be using three different systems at once for three different courses and have three different logins. Faculty development also poses a problem as trainers struggle to explain the benefits of each software package and teach them several different ways to do the same thing on three different systems.

Course management software, with all its bells and whistles, should first offer an enhanced learning environment in which students, faculty and librarians collaborate in the construction of knowledge. The pedagogy should drive the technology rather than the other way around. Having three different systems available makes it more difficult to help faculty change their traditional face-to-face teaching strategies to an on-line delivery format because software issues, that should be transparent in the process of instructional design, are brought to the forefront of the design and decision making process. Many faculty use the tool with which they have developed a comfort level and not the one that best suits their course goals and objectives. No tool will ever suit the needs of all faculty, nor is there a silver bullet technology tool that will work in all situations but it is essential for all involved parties to work together to reach a consensus on campus needs, goals and future direction of web-based teaching and learning. Two librarians have observed shrewdly that if libraries successfully establish a presence in courseware, the gain will be an increased relevance with students and strengthened relationships and collaborative ties with faculty (Shank & Dewald, 2003).

Docutek Inc.'s ERes system first enables provision of content but then enhances that content with communication tools. The recent enhancements of their virtual reference system brings the librarian and user even closer together in this context.

In an article on the future of knowledge management, Tom Davenport (1996) states that successful knowledge transfer involves neither computers nor documents but rather interactions between people. To the degree, that electronic reserve and courseware products facilitate this interaction around content will determine their success at academic institutions.

## REFERENCES

Astin, A. (1993). *What matters in college? Four critical years revisited.* San Francisco: Jossey-Bass.

Cohen, D. (2001). Course management software: The case for integrating libraries. Retrieved online, December 15, 2003, *CLIR Issues (Council on Library and Information Resources),* http://www.clir.org/pubs/issues/issues23.html.

Davenport, T. (2001). What's knowledge management got to do with IT–Part 1. Retrieved online, December 15, 2003, *Australian Computer Society Professional Development Board: Education Across the Nation,* http://www.webcom.com/quantera/Secrets.html.

Good, M. L. (1988). *New directions in applied chemistry–impact on chemical education.* Paper presented at the 10th Biennial Conference on Chemical Education, Purdue University, IN. (July 31 to August 4, 1988).

Gregoire, R., Bracewell, R., & Laferriere, T. (1996). *The contribution of new technologies to learning and teaching in elementary and secondary schools.* Quebec, Canada: Laval University and McGill University.

Hornblower, M. (1997). Great Xpectations. *Time,* 129 (23), pp. 58-68.

Jonassen, D. (1994). *Computers in schools: Mindtools for critical thinking.* State College, PA: Pennsylvania State University Press.

Kinnaman, D. E. (1993). Technology and situated learning. *Technology and Learning,* 14 (1), pp. 86.

McGrath, B. (1993). Partners in learning: Twelve ways technology changes the teacher-student relationship. Retrieved online, December 12, 2003, the *T. H. E. Journal Online* at: http://www.thejournal.com/magazine/vault/articleprintversion.cfm?aid=1982.

Russell, T. (1997). *The no significant difference phenomenon: As reported in 248 research reports, summaries and papers.* Retrieved online: http://www2.ncsu.edu/oit/nsdsplit.htm.

Shank, J., Dewald, N. (2003). Establishing our presence in courseware: Adding library services to the virtual classroom. *Information Technology and Libraries,* pp. 38-43.

Twigg, C. (1997). Putting learning on track. *Educom Review.* 30 (2), pp. 60.

West, G. B. (1999). Teaching and technology in higher education. *Adult Learning.* 10 (4) pp. 4-20.

# Docutek's ERes Electronic Reserve Software: An Evaluation

## Bud Hiller

**SUMMARY.** This paper covers selected shortcomings in Docutek's ERes system for electronic reserves. Some of the topics include an inability to customize some pages or find certain usage statistics, a high number of steps to add a document to a page, some difficulties in cross listing courses and deleting faculty accounts and a lack of options for creating non-course pages. While there are a number of items mentioned as problems within ERes, the system remains a terrific choice for institutions looking for an efficient and effective way to make faculty course materials available on the Web. Docutek is planning a major upgrade with v5 that may answer some of the questions raised in this article. *[Article copies available for a fee from The Haworth Document Delivery Service: 1-800-HAWORTH. E-mail address: <docdelivery@haworthpress.com> Website: <http://www.HaworthPress.com> © 2004 by The Haworth Press, Inc. All rights reserved.]*

**KEYWORDS.** Docutek, ERes, Bucknell University, electronic reserve

---

Bud Hiller is Technology Support Specialist, Information Services and Resources, Bucknell University, Lewisburg, PA 17837 (E-mail: dhiller@bucknell.edu).

Docutek and ERes are registered trademarks of Docutek, Inc.

[Haworth co-indexing entry note]: "Docutek's ERes Electronic Reserve Software: An Evaluation." Hiller, Bud. Co-published simultaneously in *Journal of Interlibrary Loan, Document Delivery & Electronic Reserve* (The Haworth Information Press, an imprint of The Haworth Press, Inc.) Vol. 15, No. 1, 2004, pp. 99-118; and: *A Guide to Docutek, Inc.'s ERes Software: A Way to Manage Electronic Reserves* (ed: James M. McCloskey) The Haworth Information Press, an imprint of The Haworth Press, Inc., 2004, pp. 99-118. Single or multiple copies of this article are available for a fee from The Haworth Document Delivery Service [1-800-HAWORTH, 9:00 a.m. - 5:00 p.m. (EST). E-mail address: docdelivery@haworthpress.com].

http://www.haworthpress.com/web/JILDD
© 2004 by The Haworth Press, Inc. All rights reserved.
Digital Object Identifier: 10.1300/J474v15n01_09

# ERes–SHORTCOMINGS OF A TALL PRODUCT

Oh I wish I were a Docutek programmer
For that is what I would truly like to be,
For if I were a Docutek programmer,
I'd write a system that'd be just for me.

I love to use it everyday,
And if you ask me why, I'll say
Because ERes has a way
Of making people look good each day.

(With apologies to Oscar Meyer)

## THE PROJECT

When I was asked to write this article some months ago, I thought it would be easy to come up with a thousand things that would make the list of "Items that I'd like to change if I ran the world (of Docutek)." After all, I have been working with it since 1996, as Bucknell University was one of the first schools to license the software (and it was worth every one of the 500 bucks it cost us then). Starting with version v2.0 and continuing through 3.0 and the update to 4.1, we have added, moved, deleted and linked to thousands and thousands of documents and web sites at Bucknell. During that time, dozens of professors and students have offered suggestions on how to improve the product and our service.

At the time, we were not able to immediately utilize all of this ever-so-helpful advice. We were not able to put a link to everyone's courses on the front page of every web site, or allow students the power to add documents to a page without also giving them the power to move any other items from that page. We were not able to customize professors' web pages to match the exact look of their personal home pages, or save their ERes page in its entirety to be taken with a faculty member leaving the campus. Some things are just beyond the scope of normal web service.

While we could not put some ideas into practice because of their implausibility, we did take in suggestions that made us think, "Hey, that *would* be a good idea." Unfortunately, inventing a machine that sucked in old, faint, smudged, underlined copies and spat out beautifully clear files that loaded themselves on a web page in an instant proved more difficult than we imagined. The program that we tried to write that automatically shrank file sizes to 1% of the original was buggy. The audio plug-in that would tell the user exactly where to click on the page created its own set of problems in the quiet library environment (**CLICK ON THE BLUE LOGIN LINK. GOOD JOB.**). Our contest that awarded a year's supply of low-fat avocados to the faculty member who turned in all of their electronic reserve materials first didn't catch on the way that we had hoped. Our attempt to fingerprint every student, in order to identify them as they sat down at a campus keyboard and thus do away with any sort of login at any level, did not pass muster with the ACLU. Our plan to tax students every time they asked us for the course password met with disapproval from both the College Democrats and the Conservative Club.

It is not like we did not try anything. In the end, we decided that as comments came to us about updates in our ERes service, we needed to look at either the habits and work styles of the faculty (and try to change them), or look at the software for improvements. Readers with an easily upset stomach may want to skip the next paragraph as I describe why altering the habits of professors was not an option.

I do not have anything against faculty members as individuals, but as a group, they are harder to guide than a group of 13-year old snowboarders let loose on a mountain. The concepts of complete citations, clean copies and realistic timelines seem as foggy as San Francisco Bay to many of them. A torn, blurry, faded document with underlining and highlighting, carried around for the past eleven years since some graduate school class and no citation other than the chapter title, does not a good scan make, no matter what magic, software and effort goes into it. No matter how many times this fact is emphasized, many faculty simply have too many other things to worry about. Their focus is on getting material online so they can lead a good class, not our worries about the specifics of how this is done.

We were left with creating a register of improvements in the software, ERes by Docutek. Perhaps these updates would allow us to put into effect all of the suggestions provided by faculty and students and our student employees. There was a hitch in this plan, however. We Love ERes.

## THE DILEMMA

Before embarking on a tour of ERes' shortcomings, I have to say that ERes has been terrific at Bucknell University. Not only has it been wildly successful with students and faculty as a way to make required course readings more easily available, but it is also safe to say that ERes has been one of the leading forces toward making our entire campus more technologically advanced (Hiller & Hiller, 1999). The success of ERes contributed to the positive reception granted to our merging of the library and computer services and continues to provide a solid, stable base for many courses across the campus.

In addition, the support from ERes sparkles. Whether by e-mail or phone calls, the staff at Docutek is quick to recognize the reality of the critical situations presented to them and just as quick to resolve them. It is a true credit to their staff that as the number of Docutek schools grew from a half-dozen to several hundred, the quality of their customer service did not drop.

Having sung their praises for years in online forums, in print, over e-mail and at conferences, I feel like a modern-day Benedict Arnold for writing an article that describes items that I would have liked to have seen over the years in ERes. However, no company can put everything together in the precise manner that everyone likes. No software engineer can write scripts that seem logical and clear to every person. The areas in which I would like to see improvements in ERes may not be the same that others might see because of different work flows, different types of students and faculty and different hardware setups. I want to make it clear I'm not pointing out major deficiencies in the system. I am only reflecting on various circumstances that over the years have made me sing the song that goes "I wish I were a Docutek programmer. . . ."

ERes v5, a major upgrade, is planned for 2004–perhaps some or all of these suggestions will be implemented in the coming year. Regardless, Docutek's progression over the years and our implementation of these various incarnations, makes for an interesting case study of an individual customer's desires matching up with a company's designs for the product.

## THE SUGGESTIONS–
## AT THE INDIVIDUAL COURSE LEVEL

### Passwords

Here's a little-known fact from both movie and literature history–the real reason that Col. Kurtz from Conrad's *Heart of Darkness*, played by Brando in the movie *Apocalypse Now*, fled to the most remote part of the jungle was to avoid having to create and remember one more password. Now perhaps he carried it to the extreme, but every person working at an institution of higher education edges ever nearer to that cliff of despair whenever a staff member begins a sentence with "Click on the Login button . . . " ERes is no different. To create a course you need to create an account manually and assign a password to the faculty member. At Bucknell we made the password the same as the username and then instructed the faculty member on the delightfully simple way they could change it.

Unfortunately, despite tattooing the phrase "Simply log in with your normal username as both the username and password are the same and check the box to change your password to anything you'd like" across our foreheads, we could not always overcome the double barrier that passwords present. First, the mere mention of another password causes many faculty members and students to swoon, babble and go glassy-eyed, as if Elvis himself had asked for directions to the men's room. Second, unless they have committed security sacrilege and changed every password they own to the same easily remembered word, faculty are unlikely to remember from one semester to the next what their ERes password might have been (see Figure 1).

ERes also allows for passwords at the course level. The positive angle of this option emerges when we place documents online for campus wide events–readings for an upcoming series, articles written by a speaker coming to campus, committee or parking plan reports, etc. Without needing to register for a course or contact anyone, a Bucknellian who has received an e-mail or seen a poster or story with password information can access this material, which remains mostly hidden from off-campus users.

FIGURE 1. Caution: Remembering Another Password May Be Dangerous to Your Health

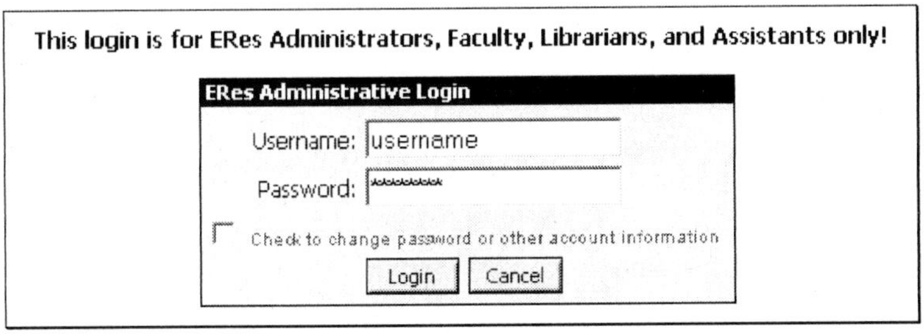

Used with permission.

This access is terrific, but it comes at a cost. Passwords must be assigned and distributed for each individual course. Some faculty give all of their courses the same password, but most provide a different password for each course. A professor teaching the standard three courses may have just added a course management password and three course passwords to an already overloaded memory storage area. Next semester, there may be three new ones coming down the pike. If there were a way for ERes to somehow offer both an ability to tie existing accounts into the Electronic reserves system while retaining the openness of campus wide offerings, many people would sleep easier at night.

## Steps to Add Materials to a Course Page

With any program, simplicity of use ranks high on the list of goals. With ERes, the very nature of the electronic beast hamstrings the program's ability to achieve this goal. Taking a document from the desktop of a computer and moving it to a web server, while simultaneously placing it on a web page, attaching restrictions on access and defining additional copyright information is not a job that can be handled in a single step. Yet many faculty who express interest in the theoretical part of the picture ("Wow, I can put my own stuff on the web.") balk at the job when confronted with the multitude of steps necessary to accomplish the goal. At that point, they are more than happy to pass the process along to us.

When it gets to us, we are in the middle of doing the same thing . . . multiplied by several hundred courses and several thousand documents. Nevertheless, there is just no getting around the process:

1. Create or obtain a document;
2. Prepare that document for going on the web;
3. Transfer that document to the web server through FTP or ERes' own file transfer program (8 clicks);
4. Add the transferred document to the course page, giving it a name and some copyright information (approximately 9 clicks);
5. Send an e-mail to the faculty member, letting her know that the file is online and my clicking finger is getting tired;
6. Repeat.

This process can either be considerably longer with more copyright or folder information to impart, or it can be shorter if you bulk-transfer a number of documents at one time. But as Page and Plant (1973) said,

"California sunlight, sweet Calcutta rain, Honolulu star bright, the song remains the same."

It really does not matter if it is 15 clicks, 13 clicks or 19 clicks. After you have done it for a few hours/days/weeks/months/years, the process gets old. Perhaps a streamlined option for your most basic workflow, adding a file-based document that you're going to transfer to the server, which would answer about 4 questions that currently require individual mouse clicks, would pay dividends over the long haul.

Similarly, the process of linking to a document on another course page, which currently involves searching for the document in pull-down menus by department, then by course, then by a list of documents, seems endless when you have a long list of departments and courses through which you need to wade.

Changing the order of documents occurs one document at a time, sliding number 14 above number 8, or number 26 above number 1. In a drag and drop world, making changes a step at a time seems awkward.

On the other hand, the steps within ERes are normally perfectly clear and straightforward. A user who pipes in likes he's on talk radio ('Hi, long time ERes user, first time ERes course manager and I have a question . . . ') usually finds that his questions would be answered if he just looked at the screens, since the steps are logical and clearly labeled. It is only when these perfectly clear, logical and straightforward steps are re-peated over and over that, the teeth begin to grind and the hands begin to clench.

### The Little Things; Admin Functions, Folders, Timing

Emerson would not have said, "Consistency is the hobgoblin of little minds" if he was instructing his friends Thoreau and Alcott how to put stuff on electronic reserve. In this line of work, having similar things appear in similar places produces happy clients and in this area, ERes sometimes fall short. For example, ERes offers a login page at the top left of the main ERes page and all of the course pages, which is good. On the main page, there is a block in the top right that offers links to news items, but if you go to the course index page, the news is moved to the bottom right and a list of ERes links (including one called Admin Functions) is now on the top right. Once you are in the course, the link on the top right is called Course Page Links and it includes a link called Page Management.

While you can essentially do the same thing from any of these links, the fact that the names change and their location differs on various pages just provides another opportunity for faculty to become confused. Faculty do not need extra opportunities. Users do not often appreciate variety as much as they appreciate being able to look in the same place for a link with the same name when they are trying to figure out how to add a document or make a link on their page.

In the early days, folders did not exist on ERes. Every item stood on its own. As the need became apparent for this additional organizational step, folders came into being, but only at a single level. With any feature in any software, the introduction of a single level immediately sparks a cry for more and better levels.

This has its negatives. As an example, look at the multitudes of options available in a complex program such as Photoshop. It contains dozens of submenus and fly-out menus, but at a certain point, this becomes counterproductive. Users can no longer find the steps to do what they want to do as they wade through the morass of options.

Therefore, ERes users find that they have only a few options about how course page items might be organized and presented on a page. There is only a single folder level. Course pages look the same as all the other course pages.

While ERes offers the option to restrict the visibility of a document (placing it online on Tuesday, for example, but making it invisible until Friday), there is not any way to make it visible at any time other than 12:01 a.m. on that date (see Figure 2).

Other than live chat and a discussion board, ERes pages offer course documents or links, but not a lot of other services.

These limited options mean that some faculty, used to being able to fine-tune or customize their class offerings to the nth degree, find that ERes restricts their ability to present their material in the most effective manner. On the other hand, probably a far greater number of faculty appreciate that their primary goal of getting materials online for their students is made easier to achieve because ERes offers an efficient, uncluttered gateway. At the service desk, I try to see both sides. However, on this matter, I tend to agree with comedian Steven Wright, who put it best when he said, "You can't have everything–where would you put it?"

### Cross Listing

I do not remember courses that were triplets, listed as ANBE370/670, BIOL368/668 and PSYC316/616, but we have courses like that at Bucknell. Since students get stuck in certain patterns, searching for courses all of the time rather than by departments or professors when that

FIGURE 2. Faculty Frequently Want to Control the Specific Time as Well as the Day That a Document Becomes Visible

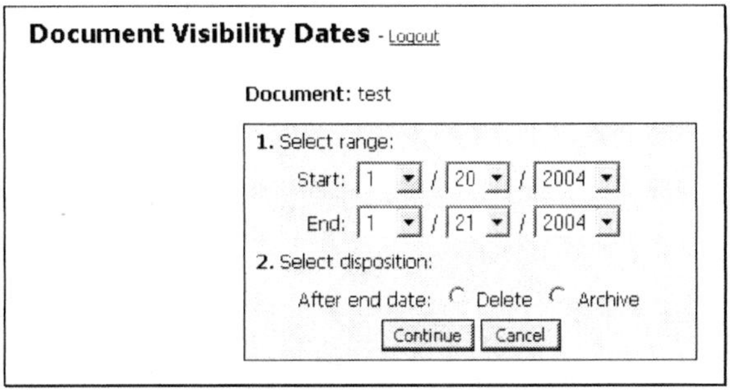

**Document Visibility Dates** - Logout

Document: test

1. Select range:

Start: 1 / 20 / 2004

End: 1 / 21 / 2004

2. Select disposition:

After end date: ⊙ Delete ⊙ Archive

Continue | Cancel

Used with permission.

option exists, we tend to try to make it possible for a student to search for any one of these courses and find the same material (see Figure 3).

ERes does not make that easy. The ideal location in the workflow for creating a course that is cross-listed would be right when the course is created. Instead, the process within ERes starts with creating one course, such as Animal Behavior 370. Since graduate students also take this course, we need to cross list it with ANBE670, because that is the course they will be searching for in the course index. Similarly, the students taking it as Biology or Psychology courses will be looking for it in those fields because they're not aware of all of the cross listings involved.

For the ERes administrator, this means going through the cross listing steps five times for this course. After clicking on the Admin Functions>Documents>Cross listing Management, the next step is to scroll through 96 pages of courses to find the one to which you want to link. Once you have done this once, you need to do it repeatedly. Ideally, this entire sequence could take place during the Create-a-Course part of the show, with a set of steps that would allow for the cross listing to be included right from the beginning. The create-a-course process asks for the course number and department–perhaps that could simply be expanded to ask for *all* of the course numbers and departments that will apply to this course.

## *Modify Course Page Info*

This section seems so innocuous, so understated and yet, *Modify page info* contains the meat of the entire course. Everything else might

FIGURE 3. The Cross Listing Page Is a Bit Confusing

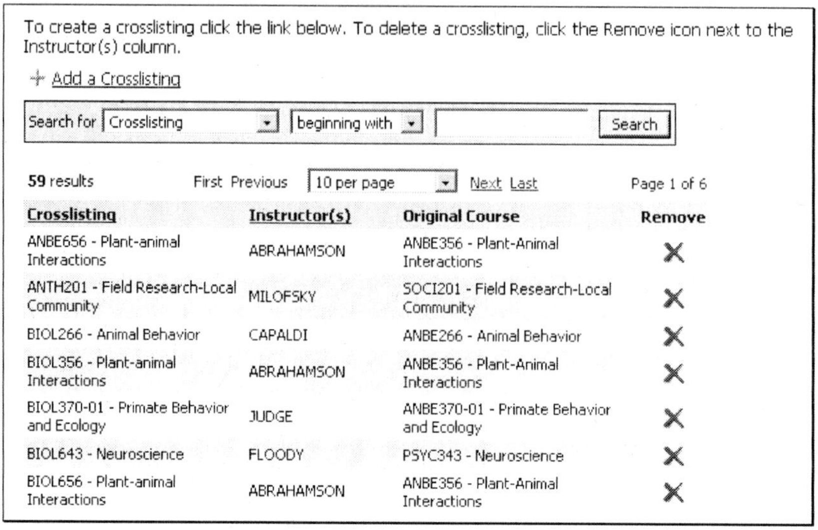

To create a crosslisting click the link below. To delete a crosslisting, click the Remove icon next to the Instructor(s) column.

✛ Add a Crosslisting

Search for | Crosslisting  ▾ | | beginning with ▾ | | | | Search |

| **59** results | First  Previous | 10 per page  ▾ | Next  Last | Page 1 of 6 |

| **Crosslisting** | **Instructor(s)** | **Original Course** | **Remove** |
| --- | --- | --- | --- |
| ANBE656 - Plant-animal Interactions | ABRAHAMSON | ANBE356 - Plant-Animal Interactions | ✗ |
| ANTH201 - Field Research-Local Community | MILOFSKY | SOCI201 - Field Research-Local Community | ✗ |
| BIOL266 - Animal Behavior | CAPALDI | ANBE266 - Animal Behavior | ✗ |
| BIOL356 - Plant-animal Interactions | ABRAHAMSON | ANBE356 - Plant-Animal Interactions | ✗ |
| BIOL370-01 - Primate Behavior and Ecology | JUDGE | ANBE370-01 - Primate Behavior and Ecology | ✗ |
| BIOL643 - Neuroscience | FLOODY | PSYC343 - Neuroscience | ✗ |
| BIOL656 - Plant-animal Interactions | ABRAHAMSON | ANBE356 - Plant-Animal Interactions | ✗ |

Used with permission.

as well be an appetizer. The course page info in Figure 4 includes the instructor, the course name and number, the password, etc. Yet, it occupies the same space and font as *View/Purge staging directory . . .* one of your lesser-known Administrative Functions. *Modify page info* is an essential. Why can't it stand out more from, say, the less glamorous *Discussion board functions*?

The *Modify page info* function could be made to stand out a bit more. When faculty have questions about changing their password or the semester in which their course is offered, they deserve to have a clear shot right to the head guy. If it makes *Modify page info* happy, we will let him keep his name, but it should be more prominent. We would love to move him out of his old neighborhood into something with a little more distinction.

## THE SUGGESTIONS–AT THE SYSTEM LEVEL

### Customize

Programmers walk the line between allowing customization and restricting access. On the one hand, some clients might have the skills and

FIGURE 4. Now, Which One of These Options Contains the Course Name, Instructor, Semester and Password?

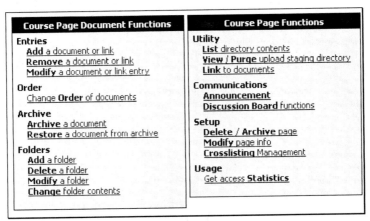

Used with permission.

desire to do their own HTML code and scripts, changing the look and the feel of their Docutek system. Some clients, luxuriating in the whole concept of being able to offer Electronic reserves without any knowledge of html, prefer to leave the entire system setup to the professionals. On the other hand, once you allow unlimited access to the code and scripts that define ERes, you open yourself to the possibility that the system may not work properly, required folders and files may be moved or deleted, the system may become unstable and users may not recognize the interface.

Overall, for its designated purpose of easily and quickly getting materials online for students, the standard design of ERes is functional, efficient and clear. After looking at screen after screen for years and years, some faculty (and some reserves professionals) yearn for some options–color schemes, navigation schemes, font choices, etc. The argument for uniformity in presentation is strong. Any student looking at any ERes course at Bucknell University knows exactly where to click to search for a course, how to access the chat room, where to look for course information and how to get back to the index. Allowing for some customization may leave some students or faculty dazed and confused, wandering the site, trying desperately to get back to where they thought they were going. For some faculty members, options equal opportunities, with the possibility of presenting material in a better way.

At the system level, this inability to customize hampers the effectiveness of the reserve staff. For example, staff cannot change the main logo on their ERes pages, that needs to be done by Docutek. The information on the right side of the page, which includes ERes links and university links, cannot be altered at the ERes site–it also needs to be done by Docutek. The links on the main home page are changeable but the style and design of the entire system remain intact. Whereas this sense of conformity makes the system work well at each individual school, staff with experience should be provided with the means to make changes in these areas if they'd like, with the accompanying added benefit of freeing up the Docutek programmers' time.

### Statistics

Before elaborating on this topic, here is a brief history lesson: system-wide statistics in v4 work tremendously better than system-wide statistics in v2. Originally, the code used to track stats in ERes went through logs that became increasingly unwieldy, so accessing statistics became the type of thing that you would click on before you went to lunch and hope it was ready by the time you got back. The current system offers a much more effective interface with results that appear on-screen within a matter of moments, so users can zip around compiling what they need instantly.

Unfortunately, faculty and administrators in libraries love usage statistics of all kinds. Who is asking for what? How many times do they ask? What is offered? When do they ask for it? How is that different from last year? How many courses are online right now and how many documents are available? ERes can provide some of this information, but not all of them.

For example, instructors can quickly check their course usage statistics to find out how many times a document has been accessed within a certain period, or how many times their course page has been accessed. There is not any way to determine who logged in to access material, but that is usually fine. Document access suffices for the majority of faculty. System-wide, there is not any way to determine the number of times documents have been accessed in a given period. Administrators can check only for ERes home page views, course page views, or course page views by course (also within a given period of time). To check on the numbers for document access system-wide for a semester, the administrator would be reduced to finding info out from each of hundreds of courses and adding them all up. Obviously, this is not viable and

'ERes home page views' does not often surface as a needed statistic. Course page views can provide a piece of the puzzle, but a single user might access the course one time and print out seventeen different documents or access the course seventeen different times to look for a single document each time. This leaves the accuracy of this statistic questionable at best (seee Figure 5).

In addition, ERes administrators frequently track the number of active ERes courses for semesters, but the system does not offer a way to look for Active ERes courses between Jan-May 2002 or any time in the past. Instead, what is available is only the number of current active courses or current archived courses. If asked to document the number of courses online this year as compared to last, the administrator could guess ('it appears to be slightly higher this semester'), or use their imagination ('we have exactly 12% more courses online this semester'), or work off of a list of active courses that they had remembered to print out at the start and end of every semester for the past seven years.

## SYSTEM-WIDE BLOCK

It is a wonderful thing to see the eyes of the faculty light up like a kid who sees snow and realizes school is cancelled. This happens at the moment that the faculty realizes how easy it is to add documents and links, move items, change the name of something, add announcements and

FIGURE 5. This Is a Summary, but It Does Not Give You Many Options for Snapshots from the Past

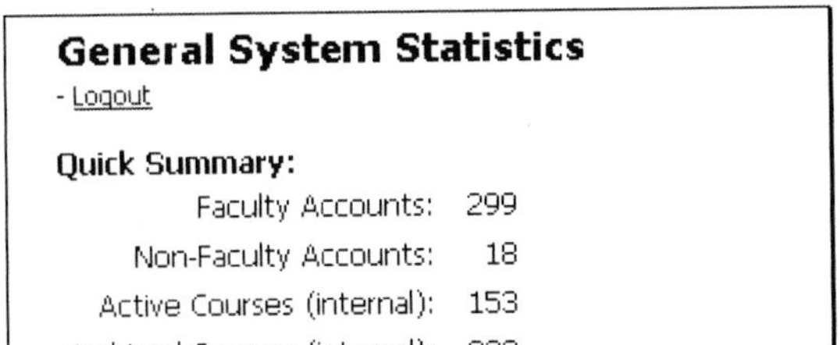

**General System Statistics**

- Logout

**Quick Summary:**

| | |
|---|---|
| Faculty Accounts: | 299 |
| Non-Faculty Accounts: | 18 |
| Active Courses (internal): | 153 |
| Archived Courses (internal): | 822 |
| External Courses: | 0 |
| Copyright Documents: | 6363 |
| Non-Copyright Documents: | 8350 |

Used with permission.

use the discussion boards in ERes. "You mean I can do all this myself?" From home, or my office, or anywhere? That is all there is to it?" they ask. "Yes," we reply, "and you can also create your own courses, archive them and restore them, link to your Blackboard course . . . you can do a lot of things."

Well, this aspect of ERes attracts faculty like beer attracts slugs, but when a gardener uses this trick, the slug suffers mightily in the long run. When the ERes administrator totally controls what goes online in ERes, the content absolutely falls into acceptable areas. The format will be one that works for all computers, the copyright guidelines will be considered and the file size will be small enough to both load and print quickly. Once the total control of this content leaves the hands of this person, who has both the training and the experience to understand why a 4mb Word Perfect® version of an entire copyrighted piece might not work on ERes, all hell breaks loose. Faculty complain that ERes is broken, the tech desk gets calls from students who can't access their material and the labs staff finds that print queues across campus have slowed to a crawl.

There is a tradeoff with ERes and relinquishing management of the system. In my experience, the file size of the document most frequently affects the ability of the students to access the documents, either by taking an inordinately long time to download or more often taking forever to print on smaller, older printers with less memory. Without giving up the option to allow faculty members to place their own material online, which they absolutely love, ERes could elaborate on their current system-wide controls for placing documents online. Currently, there is an option for document control, but it is all–encompassing–ON or OFF. Perhaps an option for adjusting the limit of the file size of the documents that are going online for a class would prevent some of the misadventures that ensue when a faculty member blindly puts an 8mb movie clip online (see Figure 6).

### Chat Program; Virtual Reference Librarian (VRL)

Bucknell University uses a chat program to provide online instant help at both the reference and tech desks. For a time, we tried to use the VRL function from within ERes and found it just too limiting. As with all aspects of ERes, some of the initial attractions included its low cost and its outstanding customer support, which included the ability to customize features exactly for our campus. We tested the chat program for

FIGURE 6. An Additional Option Here to Limit the Size of the Documents Unless Approved Would Be Helpful

| **System Configuration** - Logout | |
| --- | --- |
| Document control: | off ▼ |
| Document approval contacts (email): | |

Used with permission.

about 6 months on a trial basis in 2001 and despite ongoing efforts by Docutek to make it work for us; we eventually passed on it in order to utilize a different program.

Our biggest concerns included some items that Docutek could not control, such as the fact that we were looking for software that would work at both the tech and reference desks, not just at the reference desk. In addition, we needed software that could be mounted on our central server, not just within the ERes window, as we did not feel that the interface of going through ERes in order to ask a reference or tech-related question worked.

Since the time we researched and purchased an online chat service, Docutek has updated and improved their VRL program and some reference areas have found it to be the ideal solution for their needs (Behm, 2003). At the time, VRL worked well for a limited purpose from within ERes, but our needs called for a more generic chat program that could be placed anywhere on our combined library-computer services site.

### Deleting a Faculty Member and Bulk-Delete/Bulk-Archive

Faculty members come and go at a university–occasionally there is a tenure decision gone sour, but more frequently, there are sabbaticals, leaves, one-year replacements, short-term hires, adjuncts, etc. As a result, ERes staffs find themselves in a position where they regularly create and delete faculty accounts and courses.

Creating an account could not be more straightforward. When I've talked to faculty on the phone about how to go about getting set up on ERes, they're amazed when I tell them I've created their account during the conversation and they can begin working with their login RIGHT NOW. Courses are created, materials are added and alas, the faculty member's pleasant sojourn at Bucknell comes to an end. Unfortunately, deleting that same account takes considerably more clicks.

An account cannot be deleted if that account still has courses in the system. That is fair, as is the notification from ERes that provides links to those same courses, which must first be expunged before the account can follow suit. Unfortunately, ERes offers nary a simple method of doing this. While it is possible to:

- Click on one of the courses;
- Accept the *Copyright Agreement*;
- Go to *Page Management*;
- Click on *Delete/Archive Page*;
- Choose *Delete Page*, with the understanding that all of the documents online for this course will also disappear;
- Repeat this process for each of the several courses that a faculty member may have on ERes;
- Finally, delete the account.

This is not the ideal process. When I have trained one of our students in this process, their eyes glaze over halfway through and their next question, if they are still awake, is "Isn't there another way to do this?"

Well, there is and it is called Bulk-Delete, but it has its own drawbacks, which reappear in the Bulk-Archive option. An administrator can go into *Administrative Functions>Course Web Sites* and choose to Bulk-Restore courses that have been archived. A single long list of courses appears with a check box next to each item. Simply put a check next to each of the several courses that you would like to restore, click the *Continue* button and those courses have moved from archived to active status. All is well and right in the world and the administrator can move on their next task, satisfied and content.

To Bulk-Archive or Bulk-Delete those same courses, a different interface appears. The courses appear in groups of 10 and while it is possible to check multiple courses among those 10 and carry out the Bulk-Archive or Bulk-Delete functions, courses for the same faculty member often fall on various pages among the hundred or so that exist on the system. For example, our current system has 96 pages worth of courses if you click on the Bulk-Delete option. If I need to clear several Management classes, one Foundation Seminar and a Capstone course for a single faculty member, I can sort the files by instructor or course name, but I still need to advance through the pages one at a time in order to clear multiple courses. If I check the box to delete a course on one page, then advance to the next page and click 2 more, then click on *Continue*, only the last 2 boxes are deleted and I'll need to go back and do the first one a second time. The Bulk-Archive feature, which is interestingly the exact counterpart of the Bulk-Restore option, works the same way as Bulk-Delete (see Figure 7).

The result is that deleting an account with multiple courses takes an inversely high number of clicks and steps than creating that same account. Library staff the world over would rejoice in a simple feature that offered a choice like this when the Delete Account button is pressed (see Figure 8).

### *Non-Course Page*

ERes offers an option for presenting material to the campus that other course management software systems do not. Service committees, guest speakers, campus-wide reports, etc., can reside on ERes and with an e-mail announcement, that includes a password, course owners can let the university community know what exists and how to access it. This option works beautifully in many respects. Course owners do not

FIGURE 7. Ninety-Eight Pages Like This to Wade Through . . . ERes v5 Should Greatly Improve This Feature

**Bulk Delete Course Pages** - Logout

**975** matches    Go to page: First Previous Next Last    Page 1 of 98

| Delete | Course Number | Modified | Instructor | Course | Term |
|---|---|---|---|---|---|
| ☐ | ANBE266 | 1/15/2004 1:32:19 AM | Elizabeth Capaldi | Animal Behavior | Fall03 |
| ☐ | ANBE266 | 1/15/2004 1:38:18 AM | Peter Judge | Animal Behavior | Fall00 |
| ☐ | ANBE354 | 4/17/2003 3:32:08 PM | Elizabeth Capaldi | Tropical Ecology | Spring03 |
| ☐ | ANBE355 | 11/20/2003 3:34:33 PM | Elizabeth Capaldi | Behavior of Social Insects | Fall03 |
| ☐ | ANBE356 | 12/4/2003 12:34:12 PM | Warren Abrahamson | Plant-Animal Interactions | Fall03 |
| ☐ | ANBE370-01 | 11/8/2002 11:34:23 AM | Peter Judge | Primate Behavior and Ecology | Fall02 |
| ☐ | ANTH109 | 1/19/2004 11:50:18 AM | Edmund Q Searles | Introduction to Cultural Anthropology | Spring04 |
| ☐ | ANTH109-01 | 1/16/2002 9:12:29 AM | Susan Reed | Cultural Anthropology | Spring02 |
| ☐ | ANTH109-01 | 9/3/2003 10:13:56 AM | Michelle Johnson | Cultural Anthropology | Fall03 |
| ☐ | ANTH109-02 | 1/16/2004 2:07:29 AM | Tom Greaves | Introduction to Anthropology | Spring04 |

Continue    Check All Uncheck All

Used with permission.

FIGURE 8. A Single Button Like This Would Certainly Come in Handy When You Want to Delete an Account

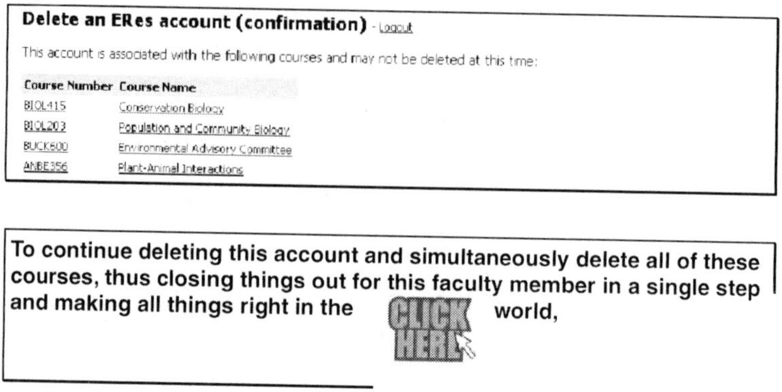

**Delete an ERes account (confirmation)** - Logout

This account is associated with the following courses and may not be deleted at this time:

| Course Number | Course Name |
|---|---|
| BIOL415 | Conservation Biology |
| BIOL203 | Population and Community Biology |
| BUCK800 | Environmental Advisory Committee |
| AhBE356 | Plant-Animal Interactions |

To continue deleting this account and simultaneously delete all of these courses, thus closing things out for this faculty member in a single step and making all things right in the **CLICK HERE** world,

Used with permission.

have to register users in order for them to look at documents or links. For something like a humanities seminar, where the organizer wants to make a few of the upcoming speaker's writings available to the campus before the visit, ERes fits the bill.

Unfortunately, to make the transition from paper to the web seamless, the administrator has to jump through a few hoops. The first is that a class must be created and an instructor provided. Naturally, a committee report or parking plan rarely meets the definition of a class and while the person who invited a speaker to campus may be a faculty member, in this instance they cannot be considered an instructor. Therefore, the ERes administrator fudges things. First, he assigns a random department and course number (at Bucknell, we created a department called Bucknell University for this purpose, so these type of repositories are BUCK501, BUCK515, etc.). Next, he gives the fictitious "course" a course name like "Abraham Lincoln, guest speaker" and proceeds exactly like this were a normal class.

To ease the process of finding these non-course materials, we created an image in the top right corner of every ERes page (see Figure 9). The link goes to a page that lists all of our online courses in the departments of Bucknell University and users can quickly scan the roll and click on their link. Alternatively, users could click on a link in an e-mail or on a web page that goes directly to the advertised "course."

FIGURE 9. The Graphic in the Top Right Corner Goes to a Page with Links to All of Our Non-Course Materials

Used with permission.

Obviously, the flaw in this system rears its ugly head when the faculty member says, "Well, it's not a course, so it doesn't belong to a department and I'm not the instructor." Hmm, yes, but those are the restraints under which we work, we say and ask them to think back to the mid '70s, when magician Doug Henning would say that everything was an "Illusion." Try not to look at BUCK509, with a course instructor, course name and semester as a course–instead, think of it as just an Illusion of a course.

Q. When is a door not a door?
A. When it is ajar.

Q. When is a course not a course?
A. When it is defined as non-course material and added to the Department of Bucknell University.

The punch line does not quite grab you the same way, but the workaround leaves us with no other choice. Perhaps ERes could exploit its initial advantage of being able to offer material to the campus by creating a specific non-course type of course that better meets the needs of its users.

## FINALE

In sum, this seems like many suggestions to improve a product that works exceedingly well. I liken it to the times that I have worked with faculty who will call me to discuss a problem that a specific student has been having accessing a particular document on ERes. At first, many faculty could not be convinced that this electronic thing could possibly work better than the old paper modes of distribution. After I pointed out that we had thousands of documents online in hundreds of courses, with dozens of students in each class accessing the materials at all hours of the day and night, the instructor came to realize that one student having troubles with one document did not signify a need for a major redistribution of our efforts.

I see the points in this paper the same way. The vast majority of scripts and tools within ERes offer an outstanding combination of straightforward design and rugged reliability and while there are a few things I would like to see changed, the system itself does not require a total rewrite. Of course, these are just my opinions as I drive my ERes 4.2.04 around our campus. Your mileage in your ERes system may vary.

## REFERENCES

Behm, L. M. (2003). One Library's Experience with Review and Selection of Chat Software for Reference. *Medical Reference Services Quarterly, Vol. 22(2), Summer.*

Hiller, B. & Hiller, T. B. (1999). Electronic reserves and success: where do you stop? *Journal of Interlibrary Loan, Document Delivery & Information Supply,* vol. 10, no, 2 pp. 61-75.

Plant, R., & Page, J. (1973). The song remains the same. On *Houses of the Holy.* Atlantic Records.

# Index

Page numbers followed by t indicate tables.

Access, 52-53,62,83-84,102-104
Adding materials, 104-105
Administrative functions, 105-106,
    111-117
  bulk delete/bulk archive, 113-115
  chat program, 112-113
  faculty-member deletion, 113-115
  non-course page, 115-117
  upgrade, 3
  virtual reference librarian, 112-113
Adobe Acrobat, 22
  Paper Capture, 23-24
  PDF Optimizer, 23
Advocacy, 47
American Library Association (ALA),
    8,68
Antivirus software, 18-19
Apache HTTP software, 14
Archiving, 47,113-115
ASP programming, 17
Association of Research Libraries, 13
AT&T, problems with, 49
Automatic document feeders, 21

Backup, 47
Bandwidth sensitivity, 46
Blackboard, 76,78-81. *See also*
    Courseware
Bucknell University, 99-118
Bulk delete/bulk archive, 113-115

(University of) California-Santa
    Barbara, 7
CampusCruiser portal, 94
Card File system, 6
CD-RW storage, 52
Central processing unit (CPU), 16

Chat program, 112-113
Cleanup, 46,53
Colorado State University-Boulder,
    68,71
Colorado State University-Fort
    Collins, 68,71
Colorado State University-Pueblo,
    65-73
Commercial resources, interoperability
    with, 27-28
Configuration, server, 19-20
CONFU (Conference on Fair Use),
    45,54,68
Connectivity, 49
Consistency, 105-106
Copyright Clearance Center, 51-52
Copyright management, 27,65-73,83
  implementation, 39,51-52,54-55
  sample form, 63
  upgrade, 3-4
Course-level suggestions
  access, 102-104
  adding materials, 104-105
  administrative functions, 105-106
  consistency, 105-106
  cross-listing, 106-107
  modification of information, 107-108
  passwords, 102-104
Course pages, upgrade, 4
Courseware
  background, 76-77,88-90
  database integration, 81-85
  DocuFax case study, 77
  interoperability with, 28
  limitations of, 95
  management tool comparisons, 91-94t
  need for integration of, 78-81
  resource integration, 87-97

CPU (central processing unit), 16
Cross-listing, 106-107
Cunningham Memorial Library (Indiana
    State University), 5-9
Customizing capability, 3,108-110

Database architecture, upgrade, 3
Database integration, 81-85
Database linking, 47,48,84
Deletion, 113-115
Dial-up services, 46
Disk defragmentation, 18
Disk Law, 16
DocuFax module
    case study, 77
    implementation, 37-38
Documentation, server, 18
Docutek Information Systems, history,
    1-2,5-9
Docutek Install Team, 33-34
Docutek Users Group, 29

Electronic reserves, defined, 5-6. *See
    also* ERes
Electronic Reserves Clearinghouse, 13
E-mail, 95
ERes. *See also* Docutek, Electronic
    reserves; *Individual subtopics*
    databases and, 81-85
    evaluation, 99-118. *See also*
        Evaluation
    founder, 6
    implementation, 31-41,43-64
    resource integration, 11-30,75-85
    upgrade, 2-4
    VRL*plus* component, 14-15
ERes User Group, 13
Evaluation, 99-118
    administrative functions, 111-117
    background, 100-101
    course-level suggestions, 102-108
    dilemma of, 101-102
    system-level suggestions, 108-111

Faculty accounts, 35-36
Faculty contract, 53-54
Faculty-member deletion, 113-115
Faculty notification, 36-37
Faculty requests, 50-51
Fair use, 65-73,83. *See also* Copyright
    upgrade, 3-4
Family Educational Right to Privacy
    Act (FERPA), 68
Fax capability, 33-34. *See also*
    DocuFax module
Feedback, 46,55,58
FERPA (Family Educational Right to
    Privacy Act), 68
File size, 4

Gigahertz (Ghz), 15-16
Goal setting, 44-47

Hackers, 18-19
Hard drive crashes, 17
Hardware/software configuration, 33-36

Implementation, 31-41
    background, 32-33,44
    connectivity, 49
    copyright management,
        39,51-52,54-55
    database linking, 48
    establishing reserve procedures,
        37-38
    faculty contract, 53-54,59-60
    faculty notification, 36-37
    feedback, 46,49,55,58
    future considerations, 40-41,57
    goal setting, 44-47
    hardware/software configuration,
        33-36
    indexing issues, 48
    lessons learned, 56
    Penfield Library (SUNY Oswego)
        experience, 43-64

pilot programs, 45,47-50
reserve process, 50-53
technology challenges, 40
update tracking, 56
user statistics, 39
workflow, 38-39,49-50
Indexing issues, 48
Indiana State University, 5-9
Information literacy, 82-83
Information packets, 47
Interoperability, 24-28
  with commercial resources, 27-28
  with courseware, 28. *See also*
    Courseware; Resource
    integration
  with online public access catalog,
    25-27

Kesten, Philip R., 5-9

Linking systems, databases, 84
Linus/Unix operating system, 13-14
Logbook, server, 18

MARC records, 3-4
Megahertz (Mhz), 15
Memory, 15-16
Microsoft Access, 13
Microsoft IIS
  patches and updates, 19
  vulnerability, 14,18-19
Microsoft.NET, 4
Mirroring of hard drive, 17
Modification of information, 107-108
Moore's Law, 15-16

(University of) North Carolina-
    Wilmington, 31-41
Northern New York Library Network,
    43-64

Online public access catalog,
    interoperability, 25-27
Operating systems, 4,13-14
  Linus/Unix, 13-14
  Microsoft Internet Information
    Services, 14,18-19
  Netscape Enterprise Web Server, 14

Passwords, 34,83-84,102-104
Paul J. Gutman Library (Philadelphia
    University), 75-85
PBS WebStudy courses, 95
PDF files
  conversion to, 22
  paper capture, 23-24
  review of, 52
PDF Optimizer, 23
Penfield Library (SUNY Oswego),
    implementation experience,
    43-64. *See also*
    Implementation
Philadelphia University, 75-85
Phys_Chat, 6-7
Pilot programs, 45,47-50
Pixels, 20-21
Platform sensitivity, 46
Processing power, 15-16

RAM (Random Access Memory), 16
Reserve process
  access, 52-53,62
  faculty requests, 50-51
  processing, 51-52
Reserve request form sample, 60-62
Resolution, scanner, 20-21
Resource integration, 11-30
  background and principles, 12
  courseware, 75-85,87-97
  interoperability, 24-28
  linking systems, 84
  literature review, 12-13
  scanners and software, 20-24
  serials management, 83
  server selection, 13-20. *See also*
    Servers

Saint Louis University Libraries, 11-30
Santa Clara University, 6,7
"Save As" option, 23
Scanners, 20-24
    automatic document feeder, 21
    color, 22
    compatibility level, 23
    duplex capability, 22
    hardware selection, 20-22
    paper size, 21
    resolution, 20-21
    software selection, 22-24
    speed, 20
Security procedures, 18. *See also*
        Access; Passwords
Sensitivity
    bandwidth, 46
    platform, 46
Serials management, 83
Servers
    configuration, 19-20
    external, 4
    hardware selection, 13-20
    management, 17-20
    memory, 16-17
    operating systems, 13-14
    processing power, 15-16
    sharing applications, 17
    speed, 15-16
Sharing applications, 17
Software base, 4
Speed
    scanner, 20
    server, 15-16
State University of New York (SUNY)
        Oswego, 43-64
Statistics, 49,110-111
    implementation, 39
Storage capacity (memory), 15-16
Storage Law, 16
Syracuse University, 6

System-level suggestions, 108-111
    customization, 108-110
    usage statistics, 110-111


Technology challenges, 40


University of California-Santa
        Barbara, 7
University of North Carolina-
        Wilmington, 31-41
URLs, 51,78
User statistics. *See* Statistics


Verizon, 49
Virtual reference librarian, 112-113
Virus protection, 18-19
Visual Basic.NET, 3
VRL (virtual reference librarian),
        112-113
VRL*plus* component, 14-15


Web servers. *See* Servers
Websites, sharing with, 17
WHYY television (Philadelphia)
        WebStudy courses, 95
Widener University, 87-97
William Madison Randall Library (U.
        North Carolina), 31-41
Wolfgram Memorial Library (Widener
        University), 87-97
Workflow
    implementation, 38-39,49-50
    upgrade, 3
Worms, 18-19


Zivkovic, Slaven, 7

# BOOK ORDER FORM!

Order a copy of this book with this form or online at:
http://www.haworthpress.com/store/product.asp?sku=5495

## A Guide to Docutek, Inc.'s ERes Software
### A Way to Manage Electronic Reserves

_____ in softbound at $19.95 (ISBN: 0-7890-2783-6)
_____ in hardbound at $39.95 (ISBN: 0-7890-2782-8)

COST OF BOOKS _____

POSTAGE & HANDLING _____
US: $4.00 for first book & $1.50
for each additional book
Outside US: $5.00 for first book
& $2.00 for each additional book.

SUBTOTAL _____

In Canada: add 7% GST. _____

STATE TAX _____
CA, IL, IN, MN, NJ, NY, OH & SD residents
please add appropriate local sales tax.

FINAL TOTAL _____
If paying in Canadian funds, convert
using the current exchange rate,
UNESCO coupons welcome.

❑ BILL ME LATER:
Bill-me option is good on US/Canada/
Mexico orders only; not good to jobbers,
wholesalers, or subscription agencies.

❑ Signature _____

❑ Payment Enclosed: $ _____

❑ PLEASE CHARGE TO MY CREDIT CARD:
❑ Visa ❑ MasterCard ❑ AmEx ❑ Discover
❑ Diner's Club ❑ Eurocard ❑ JCB

Account # _____

Exp Date _____

Signature _____
(Prices in US dollars and subject to change without notice.)

PLEASE PRINT ALL INFORMATION OR ATTACH YOUR BUSINESS CARD

| | |
|---|---|
| Name | |
| Address | |
| City | State/Province | Zip/Postal Code |
| Country | |
| Tel | Fax |
| E-Mail | |

May we use your e-mail address for confirmations and other types of information? ❑ Yes ❑ No We appreciate receiving
your e-mail address. Haworth would like to e-mail special discount offers to you, as a preferred customer.
**We will never share, rent, or exchange your e-mail address.** We regard such actions as an invasion of your privacy.

Order From Your **Local Bookstore** or Directly From
**The Haworth Press, Inc.** 10 Alice Street, Binghamton, New York 13904-1580 • USA
Call Our toll-free number (1-800-429-6784) / Outside US/Canada: (607) 722-5857
Fax: 1-800-895-0582 / Outside US/Canada: (607) 771-0012
E-mail your order to us: orders@haworthpress.com

**For orders outside US and Canada,** you may wish to order through your local
sales representative, distributor, or bookseller.
For information, see http://haworthpress.com/distributors

_(Discounts are available for individual orders in US and Canada only, not booksellers/distributors.)_

**Please photocopy this form for your personal use.**
www.HaworthPress.com

BOF05